Water Sources

PRINCIPLES AND PRACTICES OF WATER SUPPLY OPERATIONS SERIES

Water Sources, Fourth Edition

Water Treatment, Fourth Edition

Water Transmission and Distribution, Fourth Edition

Water Quality, Fourth Edition

Basic Science Concepts and Applications, Fourth Edition

Water Sources

Fourth Edition

Printed in the United States of America.

Project Manager/Senior Technical Editor: Melissa Valentine
Technical Editor: Linda Bevard
Production Editor/Cover Design: Cheryl Armstrong
Production Services: Zeke Hart and Melissa Stevenson, TIPS Technical Publishing, Inc.

Disclaimer

Many of the photographs and illustrative drawings that appear in this book have been furnished through the courtesy of various product distributors and manufacturers. Any mention of trade names, commercial products, or services does not constitute endorsement or recommendation for use by the American Water Works Association or the US Environmental Protection Agency. In no event will AWWA be liable for direct, indirect, special, incidental, or consequential damages arising out of the use of information presented in this book. In particular, AWWA will not be responsible for any costs, including, but not limited to, those incurred as a result of lost revenue. In no event shall AWWA's liability exceed the amount paid for the purchase of this book.

Library of Congress Cataloging-in-Publication Data

Koch, Paul, 1960-
Water sources / by Paul Koch. — 4th ed.
p. cm. — (Principles and practices of water supply operations series)
Rev. ed. of: Water sources. 2003.
Includes bibliographical references and index.
ISBN-13: 978-1-58321-782-5 (alk. paper)
ISBN-10: 1-58321-782-7 (alk. paper)
1. Water-supply. 2. Water quality. 3. Water resources development. I. American Water Works Association. II. Water sources. III. Title.
TD390.W38 2010
628.1'1—dc22

2010016774

6666 West Quincy Avenue
Denver, CO 80235-3098
303.794.7711
www.awwa.org

Contents

Foreword

Water Sources is part one in a five-part series titled Principles and Practices of Water Supply Operations. It contains information on water supply sources, the use and conservation of water, source water quality, and source protection.

The other books in the series are

Water Treatment
Water Transmission and Distribution
Water Quality
Basic Science Concepts and Applications (a reference handbook)

References are made to the other books in the series where appropriate in the text.

The reference handbook is a companion to all four books. It contains basic reviews of mathematics, hydraulics, chemistry, and electricity needed for the problems and computations required in water supply operations. The handbook also uses examples to explain and demonstrate many specific problems.

Acknowledgments

This fourth edition of *Water Sources* was revised to include updated descriptions of water source issues and new regulations. More recent bibliographic references have also been added. The revision effort was led by Paul Koch, Ph.D., P.E. Substantial review comments were provided Chi Ho Sham and Frederick Bloetscher.

The third edition of *Water Sources* was revised to include new technology and current water supply regulations. Jane Rowan of Schnabel Engineering, Inc., and George Rest of O'Brien & Gere Engineers, Inc., provided coordination and technical review of the revision. The authors of the revision were Paul I. Welle, P.E., senior associate, Schnabel Engineering, Inc., (chapter 1); Mark H. Dunscomb, P.G., senior geologist, Schnabel Engineering, Inc. (chapter 2); John P. Harrison, P.E., Associate Engineer, Schnabel Engineering, Inc. (chapter 3); Lloyd Gronning, P.E., Parsons (chapter 4); Thomas Dumm, P.E., senior project associate, O'Brien & Gere Engineers, Inc. (chapter 5); Harish Aurora, Ph.D., P.E., senior technical associate, O'Brien & Gere Engineers, Inc. (chapter 6); and Jane O. Rowan, P.W.S., associate scientist, Schnabel Engineering, Inc. (chapter 7). Tony Wachinski, Ph.D., P.E., Pall Water Processing, and Bill Lauer, utility quality programs manager, American Water Works Association, provided overall review of the third edition.

The authors of the second edition were R. Patrick Grady, assistant editor, and Peter C. Karalekas Jr., editor, with the *Journal of the New England Water Works Association*. Reviewers of the second edition included Tom Feeley, Eugene B. Golub, William T. Harding, Stephen E. Jones, Kenneth D. Kerri, Gary B. Logsdon, and Garret P. Westerhoff. AWWA staff who reviewed the manuscript included Robert Lamson, Ed Baruth, Bruce Elms, and Steve Posavec. Harry Von Huben served as series editor.

Publication of the first edition was made possible through a grant from the US Environmental Protection Agency, Office of Drinking Water, under Grant No. T900632-01. The principal authors were Joanne Kirkpatrick and Benton C. Price, under contract with VTN Colorado, Inc.

The following individuals are credited with participating in the review of the first edition: E. Elwin Arasmith, Edward W. Bailey, James O. Bryant Jr., Earle Eagle, Clifford H. Fore, James T. Harvey, William R. Hill, Jack W. Hoffbuhr, Kenneth D. Kerri, Jack E. Layne, Ralph W. Leidholdt, L.H. Lockhart, Andrew J. Piatek Jr., Fred H. Soland, Robert K. Weir, Glenn A. Wilson, Leonard E. Wrigley, and Robert L. Wubbena.

Introduction

Supplying water to the public is one of the most important services offered to society by water supply professionals. It has grown in importance and complexity throughout the late nineteenth, twentieth, and early twenty-first centuries. With the advent of the Safe Drinking Water Act, signed by President Ford in 1974, a new era of responsibility dawned for the water utility operator.

Prior to that time, the operator's job was in large part mechanical. He or she kept the pumps running, the valves functioning, and the simple disinfection process from shutting down. The operator changed charts, washed filters, kept the buildings and grounds tidy, and handled a few basic chemicals.

Now the operator does all of those things, but with greater care and greater precision. He must still be a good mechanic, but now, if the job is to be done well, he must also have some basic knowledge of microbiology, chemistry, laboratory procedures, and public relations. The operator must be ready to explain to the community what his job consists of, why water rates seem to rise continually, and that an adequate supply of safe and palatable water is still a great bargain. The operator must reach out to the schools in the service area and teach students of all ages, both in the classroom and at the treatment plant, about the water treatment process. Most important of all, the operator must constantly be aware of public health issues. Any action she takes must be taken only if the public health has been considered. The operator has the potential to create serious health hazards by negligence or to make dangerous situations harmless by using her knowledge and skill.

As is true of any quality product, the raw material is very important. The water utility operator's raw material is the source water. Every source is different, and the operator must become the most knowledgeable, best-informed person about the source or sources he or she is working with. This is true whether the source is a single well, a number of well fields, a spring, a river, or a lake.

This book will introduce the operator to the basic information related to water sources. The operator's goal should always be to take the available source water and perform the functions necessary to supply the customers with a safe, high-quality product.

CHAPTER 1

Water Supply Hydrology

Hydrology is the study of the properties, distribution, and circulation of water and its constituents as it moves through the atmosphere, across the earth's surface, and below the earth's surface. In the context of drinking water supply operations, hydrology is primarily concerned with the factors that affect the availability and the quality of water needed to meet user demands. This chapter introduces those factors and presents some terms that are commonly used in the drinking water supply industry to describe the volume and flow of water.

The Hydrologic Cycle

The continuous circulation of water above, below, and across the surface of the earth is called the water cycle or, more commonly, the hydrologic cycle. Illustrated in Figure 1-1, the hydrologic cycle includes the following processes:

1. *Evaporation*: Water moves off land and water surfaces and into the atmosphere.
2. *Transpiration*: Water is released into the air by plants, primarily through their leaves.
3. *Advection*: Water moves with the air currents in the atmosphere.
4. *Condensation*: Water vapor in the air forms tiny droplets.
5. *Precipitation*: Water falls out of the atmosphere as rain, snow, or ice.
6. *Interception*: Some precipitation lands on vegetation and does not reach the ground.
7. *Infiltration*: Some of the rain that reaches the ground soaks into it.
8. *Subsurface flow*: Below the surface, water movement is influenced by gravity and the presence of natural barriers in the rock or soil.
9. *Runoff*: A portion of the water that reaches the ground flows toward nearby bodies of water.
10. *Channel flow*: Runoff eventually flows into small channels that feed into larger channels that carry rivers and streams.
11. *Storage*: Water is taken out of circulation because it is frozen, held in a lake aboveground, or held in an aquifer belowground.
12. *Snowmelt*: Water that has collected as snow or ice is released as liquid.

Each of these processes is explained further below.

Evaporation

Evaporation occurs when liquid water changes to an invisible gas or vapor. Through evaporation, a wet towel dries out when it is hung up, for example, and a wet driveway dries after a rain shower. Evaporation is more rapid from surfaces that have been warmed by

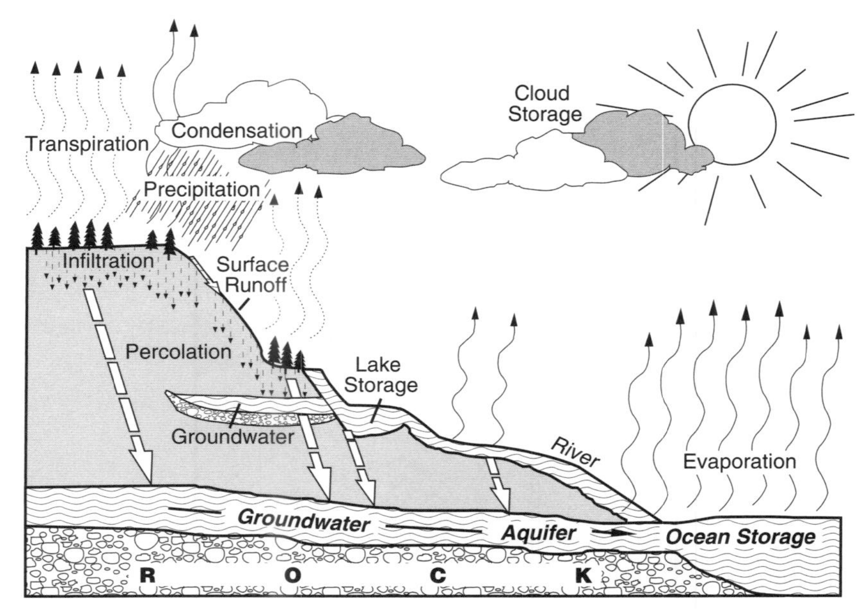

FIGURE 1-1 The hydrologic cycle

the sun or other heat sources. Across the earth, enormous amounts of water are evaporated from oceans, lakes, and streams, as well as from the land surface. The water vapor then moves with the air vertically or horizontally above the earth.

Transpiration

Transpiration also moves water into the air. Water in the soil is taken up by the roots of plants. It then moves up into the plants' leaves and is released into the air through tiny pores in the leaves called stomata.

Advection

Advection, the movement of water vapor as it is carried by air currents, redistributes water evaporated from water and land surfaces.

Condensation

Most of us have observed condensation. On a hot summer day, for example, a glass of cold liquid is quickly covered with very tiny drops of water that eventually combine and run down the glass. When conditions are right, this process of condensation happens in the atmosphere: Water vapor condenses on microscopic particles in the air and forms clouds that are made up of tiny water droplets or ice crystals.

Precipitation

As water droplets collide and combine or as ice crystals grow, they become too heavy to stay airborne and fall to the earth as drizzle, rain, hail, sleet, or snow. This process is called precipitation. The world's freshwater supply depends entirely on the various forms of precipitation.

Interception

When it first starts to rain during a storm event, the initial rainfall might not reach the ground as it encounters vegetation or other structures above ground level. The amount of rainfall that is intercepted above the ground can be substantial in wooded areas. Interception can keep relatively small storm events from increasing moisture in the ground.

Infiltration

Some of the rain or snow that falls on the ground during a storm event soaks into the soil. The movement of water through the earth's surface into the soil is called infiltration. How quickly the water moves through the soil surface depends on several factors including the type of soil, how dry it is to begin with, and whether or not a crust has formed on its surface.

Subsurface Flow

After the water penetrates into the ground, it can move in several different directions. Relatively close to the surface, water can be taken up by the roots of plants.

Another portion of the infiltrated water can be drawn back up to the surface because of the capillary action of the soil. Capillary action is the force that causes water to rise above a water surface through porous soil. The capillary rise in coarse sand is only about 5 in. (125 mm), but it may be as much as 40 in. (1,000 mm) in some silty soil.

Most of the remaining infiltrated water continues to move downward below the root zone to a water-saturated area. This downward movement of water is called percolation.

In the subsurface, water can move toward stream channels and emerge to feed the streams. This kind of water movement is what keeps water flowing in streams as the base flow between storm events.

Runoff

During a storm event, when the precipitation rate exceeds the infiltration capacity of the soil, the excess precipitation flows downhill over the land surface. This process is called overland flow or surface runoff.

Surface runoff flows along the path of least resistance and often moves toward a primary watercourse or channel.

Channel Flow

Downhill from the land surfaces where runoff is first generated, the water typically begins to flow into channels. As these channels combine, the size of channels carrying the water increases until the water is flowing in a permanent channel such as a stream or river, or until it reaches a wetland or lake. Eventually, much of the water carried in channels of various sizes and shapes reaches the oceans or inland seas (such as the Dead Sea and the Great Salt Lake).

Storage

Water does not move continuously at the same rate around the globe but may be stored for awhile and then released after a short time or a very long time. Water in the atmosphere is stored as vapor that may become visible as clouds. Surface water may be stored in ponds, natural lakes, and artificial reservoirs. Belowground, large aquifers may store water for a few years to thousands of years. At high elevations and in polar regions, water may be stored as ice and snow.

Snowmelt

Snow is a very important form of water storage. Snowmelt greatly prolongs the flow in many streams. If rainfall were the only source of water for surface water bodies, many streams would have very low flows or, in some cases, no flow during dry periods.

Many water supply systems, including most of those in the western United States and Canada, are dependent on the seasonal accumulating and melting of snow and glacial ice to provide an adequate water supply. The amount of available surface water in the spring and summer is affected by the amount of snow that accumulates in the winter and by spring temperatures. Climate warming trends can, in the short term, result in higher rates of glacial melt and an associated increase in the amount of water available downstream over several years in the near term. But over the long term, the amount of water released will decrease unless a sufficient amount of water accumulates as snow and ice each winter.

GROUNDWATER

Groundwater accumulates as the result of percolation of water down to the water table through the void spaces in the soil and cracks in rock formations. (See chapter 2 for more information on groundwater sources.) The place where groundwater accumulates is called an aquifer. An aquifer is defined as any porous water-bearing geologic formation. The size, thickness, and depth of aquifers can vary considerably. The largest aquifer in the

United States is the Ogallala Aquifer, which underlies 147,000 square miles (380,728 square kilometers) across eight states.

An easy way to visualize groundwater is to fill a glass bowl halfway with sand, as shown in Figure 1-2. If water is poured onto the sand, it infiltrates into the sand and percolates down through the voids until it reaches the watertight (impermeable) bottom of the bowl. As more water is added, the sand becomes saturated and the water surface in the sand rises. The water in this saturated sand is equivalent to groundwater in an aquifer.

When the sand in the bowl is partially saturated with water, the level of the water surface can be found by poking a hole or cutting a stream channel in the sand. As illustrated in Figure 1-3, the level of water in the channel is the same as the level of the water surface throughout the sand. This level is called the water table. The water table may also be called the top of the groundwater, the top of the zone of saturation, or the top of the aquifer.

Aquifers and Confining Beds

Groundwater occurs in two different kinds of aquifers: unconfined aquifers and confined aquifers.

Unconfined aquifers

When the upper surface of the saturated zone is free to rise and fall, the aquifer is referred to as an unconfined aquifer or water-table aquifer (Figure 1-4). Wells that are constructed to reach only an unconfined aquifer are usually referred to as water-table wells. The water

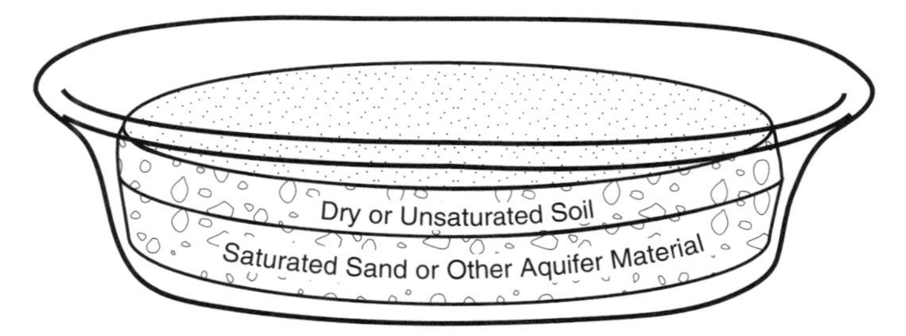

FIGURE 1-2 Simplified example of groundwater

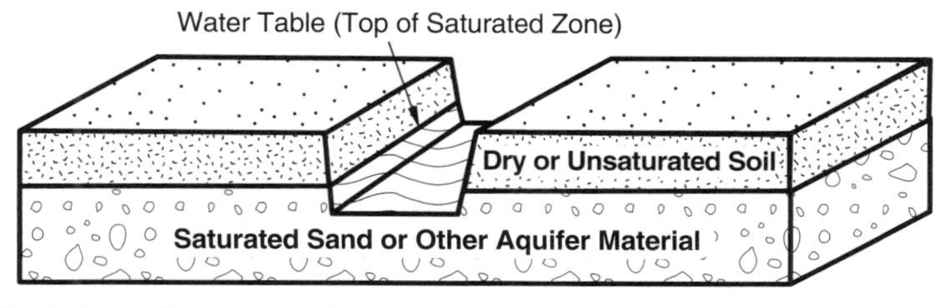

FIGURE 1-3 Illustration of a water table

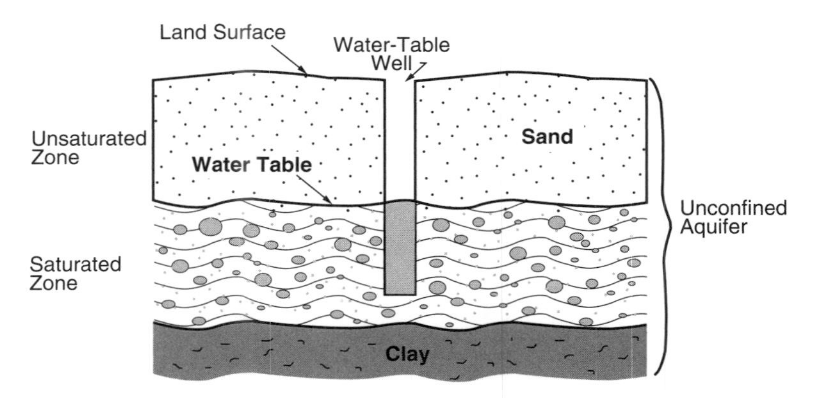

FIGURE 1-4 Cross section of an unconfined aquifer

level in these wells indicates the level of the water table of the associated aquifer. The amount of water that the water-table well can produce may vary widely as the water table rises and falls in relation to the amount of rainfall that feeds or recharges the aquifer through percolation.

Confined aquifers

A confined aquifer, or artesian aquifer, is a permeable layer that is confined between upper and lower layers that have low permeability. These layers, called confining beds (also known as aquitards or aquicludes), may consist of consolidated rock or clay. They restrict the movement of groundwater into or out of the aquifers.

As illustrated Figure 1-5, the recharge area (the location where water enters the aquifer) is at a higher elevation than the main portion of the artesian aquifer. As a result, the water is usually under pressure. Consequently, when a well is drilled through the upper confining layer, the pressure in the aquifer forces water to rise up above the confining layer. In some cases, the water rises above the ground surface, producing a flowing artesian well. In most cases, it does not rise as high as the ground surface and produces a non-flowing artesian well.

The height to which water will rise in wells located in an artesian aquifer is called the piezometric surface. This surface may be higher than the ground surface at some locations and lower than the ground surface at other points of the same aquifer. Further discussion of the piezometric surface is provided in the hydraulics section of *Basic Science Concepts and Applications*, part of this series.

There can be more than one aquifer beneath any particular point on the ground. For instance, there may be a layer of gravel, a layer of clay, another layer of sand or gravel, and another layer with low permeability. Some wells are constructed to tap several different aquifers. In locations where the water quality among aquifers varies, a well may be constructed to draw water only from the aquifer with the best quality.

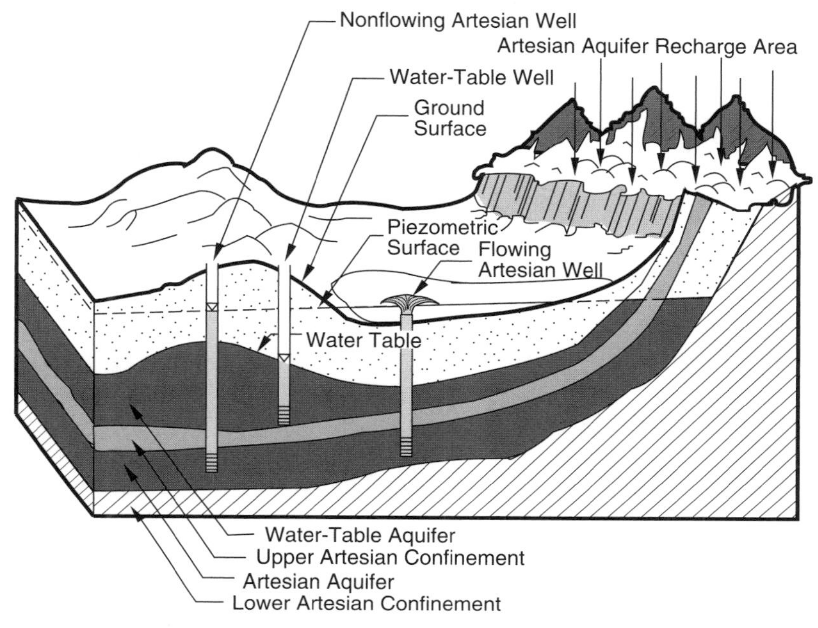

FIGURE 1-5 Cross section of a confined aquifer

Aquifer Materials

Aquifers may be made up of a variety of water-bearing materials. One characteristic that is important from the standpoint of water supply is the porosity, which represents the fraction of void space within a given amount of the material. The porosity determines how much water the material can hold. The other important characteristic is how easily water will flow through the material, a feature known as its permeability, or hydraulic conductivity. Both of these factors determine the amount of water an aquifer will yield. Figure 1-6 illustrates common aquifer materials.

Aquifers composed of material with individual grains such as sand or gravel are called unconsolidated formations. The range of sizes and arrangement of the grains in the material are important. For example, two aquifers—one composed of fine sand and one of coarse sand—may contain the same amount of water, but the water flows faster through the coarse sand. A well in the coarse sand can therefore be pumped at a higher rate, and the pumping costs will be less.

Limestone and fractured-rock formations are consolidated formations. These formations produce water from channels, fractures, and cavities in the rock. Some fractured-rock formations can produce very large quantities of water. Limestone and fractured-rock formations with rapid flows are also more vulnerable to microbiological contamination.

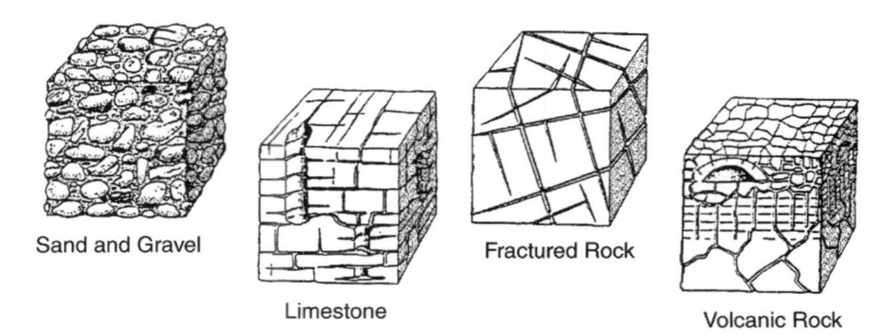

FIGURE 1-6 Common aquifer materials

Groundwater Movement

Water naturally moves downhill (downgradient) toward the lowest point. In the examples shown in Figures 1-2 and 1-3, the water will not move in any direction because the water table is flat. However, water tables are not usually flat.

Figure 1-7 illustrates how a water table might actually occur in nature. When rain falls on the watershed, some of the rain infiltrates into the soil and percolates downward to the water table. A mound of water within the aquifer is then built up above the level of the rest of the water table. The water within this mound slowly flows downgradient, increasing the level of the water table slightly and, if the water table is high enough, draining into the stream channel. In most areas, more rain occurs before the mound completely drains off, so the water table never becomes level.

The movement of groundwater is illustrated further in Figure 1-8, in which a cutaway view shows a section of the ground surface. The water table is continuous and sloped, and the groundwater moves downgradient toward the lowest point—the stream channel. In this example, the water table is above the level of the stream channel. Consequently, water will flow into the stream. But, if the water table is below the streambed, water will infiltrate from the stream to the aquifer.

Springs

Springs occur where the water table intersects the surface of the ground or where the water-bearing crevices of fractured rock come to the surface (Figure 1-9). Springs flowing out of sand or soil often come from a source that is relatively close. Springs flowing out of fractured rock can come from more distant locations, and their origin is usually difficult to determine.

Because of the difficulty in determining where spring water is coming from, springs should generally be considered contaminated until proven otherwise. Many supposed "springs" have stopped flowing after a broken underground water pipe was repaired.

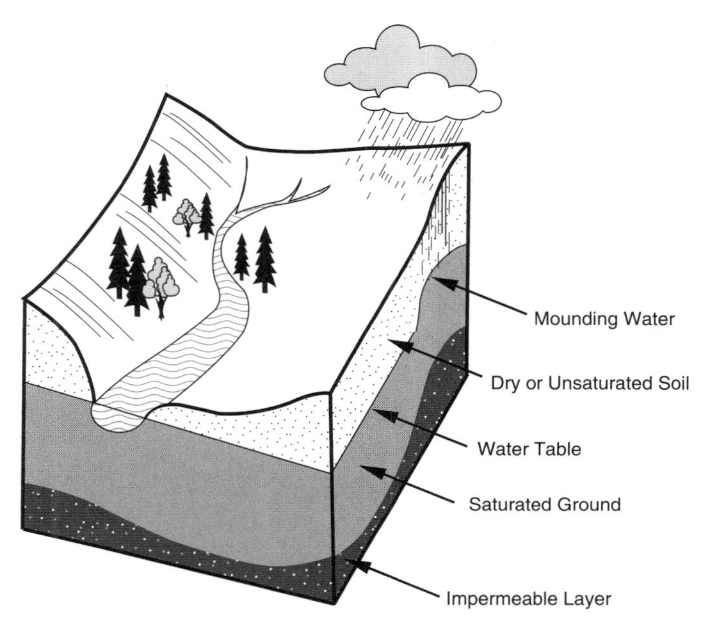

FIGURE 1-7 Formation of groundwater in nature

SURFACE WATER

Surface water in lakes and streams provides a source of water for many public water supply systems. (See chapter 3 for more information on surface water sources.) The amount of surface water available varies widely by region of the country and also by season of the year.

The quality of the surface water available for water system use also varies. Some locations are fortunate to have very clean water available; in other areas, the only available sources require extensive treatment to make the water safe and palatable for human consumption.

Chemical Constituents of Rain and Snow

Precipitation in the form of rain or snow is the source of water for most surface-water supplies. The amount of foreign material in rain and snow is minimal in comparison with the amount that the water will later pick up as it moves over land or through the ground. In general, precipitation dissolves the gases in the atmosphere as the water falls to the ground, and it also collects dust and other solid materials suspended in the air. Once on the ground, flowing water picks up soil particles and microbes and dissolves various chemicals it comes in contact with.

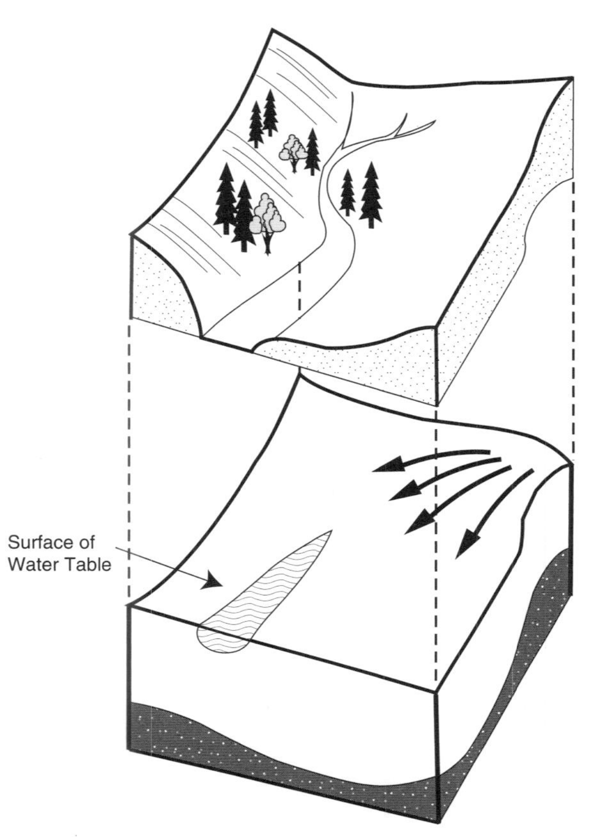

FIGURE 1-8 Groundwater movement

Atmospheric deposition is the process whereby airborne particles and gases are deposited on the earth's surface. Because the particles and gases in the atmosphere are from natural sources such as forest fires, volcanoes, and oceanic salt and from human-induced windblown soil and materials released into the air from combustion, industrial processes, motor vehicles, and other sources, there is considerable variation in the constituents of rain and snow. In general, precipitation without the impact of human activities is chemically very soft, is low in total solids and alkalinity, has a pH slightly below neutral (pH 7), and is corrosive to most metals.

There is considerable variability in the constituents of precipitation, depending on local conditions. The National Atmospheric Deposition Program monitors the chemistry of precipitation at more than 200 sites nationwide. Annual average concentrations of the constituents of precipitation are shown in Table 1-1.

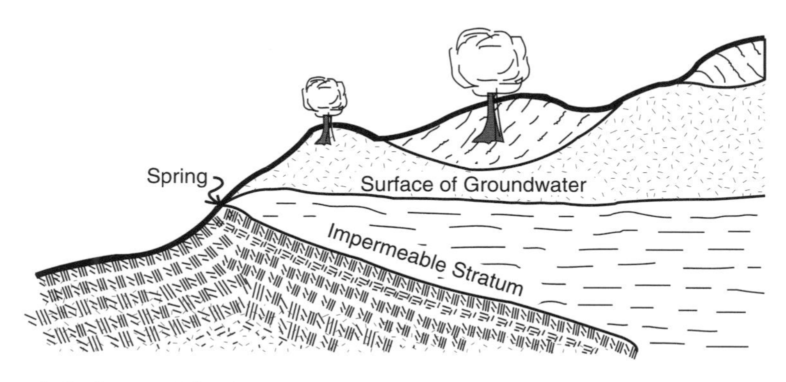

A. Spring resulting from an outcropping of impermeable material

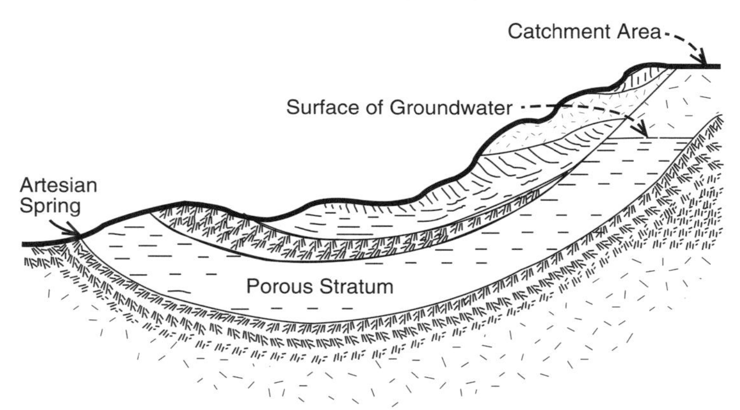

B. Spring resulting from a recharged porous stratum

FIGURE 1-9 Common types of springs

Constituents Added During Runoff

Precipitation that falls on the ground surface and does not infiltrate into the soil or evaporate into the air travels over the land surface to a surface water body. As it travels, a variety of materials may be dissolved into the water or taken into suspension. Consequently, the type and amount of various constituents reflect both the surface characteristics and the geology of the area.

The presence of soluble formations near the surface (such as gypsum, rock salt, and limestone) has a marked effect on the chemical characteristics of surface water. On the other hand, where the geological formations are less soluble, as is the case with sandstone or granite, the composition of the surface water in the area tends to be more similar to the composition of local rain.

TABLE 1-1 Annual average concentrations of the constituents of precipitation

Calcium:	0.02–1.27 mg/L as calcium (Ca ions)
Magnesium:	4–105 µg/L as magnesium (Mg ions)
Sodium:	15–980 µg/L as sodium (Na ions)
Ammonium:	0.01–0.95 mg/L as ammonium (NH_4 ions)
Chloride:	0.03–1.639 mg/L as chloride (Cl ions)
Sulfate:	0.1–2.6 mg/L as sulfate (SO ions)
Nitrate:	0.2–2.2 mg/L as nitrate (NO_3 ions)
pH:	4.3–6.3
Mercury:	3.6–19.4 ng/L as mercury (Hg)

Source: National Atmospheric Deposition Program (NADP) 2000 Annual Summary. 2001. Additional data are available at the NADP Internet site: http://nadp.sws.uiuc.edu.

If water flowing over the ground surface is exposed to decomposing organic matter in the soil, the carbon dioxide level of the water is increased. An increase in carbon dioxide causes the formation of additional carbonic acid and lowers the pH of the water. Consequently, the water becomes more corrosive. This in turn increases the amount of mineral matter dissolved by the water.

Groundwater Augmentation of Surface Water

Figure 1-10 illustrates how streamflow is augmented by groundwater. If the water table adjacent to the stream is at a higher level than the water surface of the stream, water from the water table will flow to the stream (a gaining stream). This is referred to as groundwater discharge. Conversely, when the water table is below the stream surface, water from the stream infiltrates into the ground (a losing stream). The recharge of groundwater from a stream is illustrated in Figure 1-11.

Surface Runoff

The drainage basin or watershed is the land surface that contributes water to area streams and water bodies. The amount and flow rate of surface runoff through the drainage basin are highly variable, depending on both natural conditions and human influences. In some

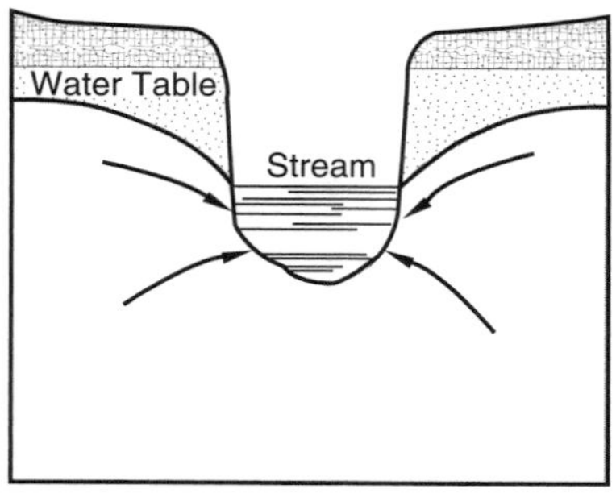

FIGURE 1-10 Surface flow from groundwater seepage

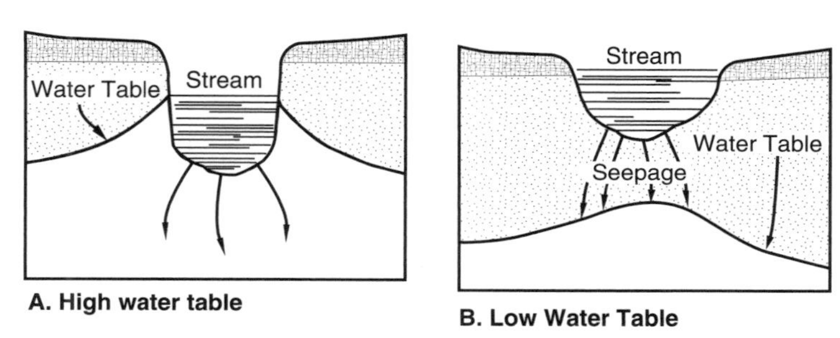

FIGURE 1-11 Streamflow recharging groundwater

cases, water is held on the surface for a relatively long time. This is generally good from a water resources standpoint because it allows more water to infiltrate into the ground and recharge aquifers. If water runs off slowly, it also generally causes less erosion and creates less flooding. At the same time, the longer the water is in contact with the soil, the greater the mineral content of the water will be. Surface water running quickly off land may be expected to have all of the opposite effects.

As shown in Figure 1-12, a drainage basin or watershed is surrounded by high ground—a divide—that separates one watershed from another. The land area of the drainage basin or watershed is sloped toward a watercourse.

Watercourses

Typical natural watercourses include brooks, creeks, streams, and rivers. There are also many human-made features that are constructed to hasten the flow of surface water or to divert it in a direction different than it would flow under natural conditions. Such structures include ditches, channels, canals, aqueducts, conduits, tunnels, and storm sewers.

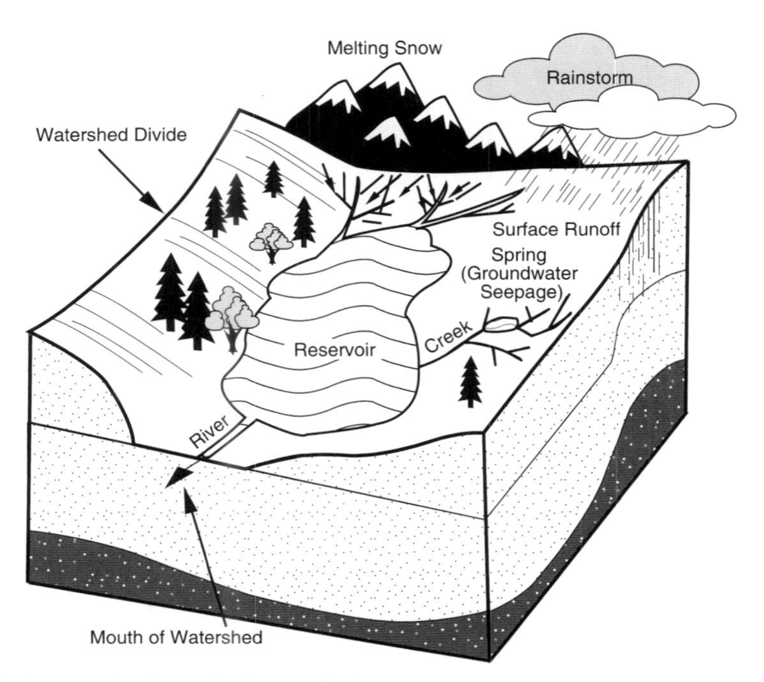

FIGURE 1-12 Schematic of a typical watershed

Impoundments

An impoundment is a surface-water storage basin. It can be a natural basin such as a wetland, pond, lake, or reservoir constructed to serve some specific water need. Reservoirs range from small basins constructed by farmers for watering livestock to massive reservoirs that store water for public water systems, recreation, and/or other uses. Impoundments can significantly affect the flow and quality of water in streams downstream of the reservoir.

VOLUME AND FLOW

The branch of science that deals with the physics of fluids at rest and in motion is known as hydraulics. A water system can be described using a variety of hydraulic measurements.

Many of the measuring instruments available to a water system operator allow the operator to simply look at a gauge and take a reading to get a flow or volume measurement without performing further calculations. An example is the water meter. Both the small residential type and the type used in a pump station or a treatment plant are designed so that the volume of water is measured continuously and the total amount of water passing through the meter is indicated in gallons (liters) or cubic feet (cubic meters).

Another example is a flowmeter, which converts the passage of water through it to a rate of flow. A single instrument frequently makes these two measurements.

Where continuous measurements need to be taken, electronic data loggers can store the measurement data until they are manually transferred to a computer. Or, if the measuring equipment can be connected to a communication network, the measurement data can be transferred automatically to wherever they need to be evaluated.

Several terms relate to measuring water. The following paragraphs discuss these terms.

Volume

Volume is the measurement of a quantity of water. In English units, volume is most often expressed in gallons, millions of gallons, cubic feet, or acre-feet. In metric units, it is measured in liters, millions of liters, cubic meters, or hectare-meters. A volume measurement may describe such things as the amount of water contained in a reservoir or the quantity of water that has passed through a treatment plant.

Flow Rate

The measurement of the volume of water passing by a point over a period of time is called the flow rate. Following are some of the units most frequently used for expressing flow rate:

- Gallons per minute (gpm) or liters per minute (L/min)
- Gallons per hour (gph) or liters per hour (L/hr)
- Gallons per day (gpd) or liters per day (L/d)
- Million gallons per day (mgd) or megaliters per day (ML/d)
- Cubic feet per second (ft^3/sec or cfs) or cubic meters per second (m^3/sec)

Instantaneous Flow Rate

The rate at which water passes by a point at any instant is called the instantaneous flow rate. The instantaneous flow rate through a water plant, for instance, varies constantly throughout the day.

Average Flow Rate

Dividing the total volume of water passing a point by a length of time provides an average flow rate. For instance, most water system reports show the average use for each day in gallons per day (gpd) or liters per day (L/d).

Some other average flow rates that are frequently computed for water supply system operations are the following:

- Annual average daily flow
- Peak-hour demand
- Peak-day demand
- Minimum-day demand
- Peak-month demand
- Minimum-month demand

Table 1-2 summarizes measurements of water use that are commonly used in the public water supply industry.

SELECTED SUPPLEMENTARY READINGS

Basic Ground-water Hydrology. 1983. Pub. WS 2220. Denver, Colo.: US Geological Survey.

Black, P.E. 1996. *Watershed Hydrology,* 2nd ed. Chelsea, Mich.: Sleeping Bear Press, Inc.

Ground Water and Surface Water; a Single Resource. 1998. US Geological Survey Circular 1139. Denver, Colo.: US Geological Survey.

Inside Rain, A Look at the National Atmospheric Deposition Program. 1999. National Atmospheric Deposition Program. Brochure 1999-01b (revised). Illinois State Water Survey. Available: http://nadp.sws.uiuc.edu.

Lehr, J.H., and J. Keeley, eds. 2005. *Water Encyclopedia.* New York: John Wiley.

Manual M32. Computer Modeling of Water Distribution Systems. 2005. Denver, Colo.: American Water Works Association.

Manual of Instruction for Water Treatment Plant Operators. 1989. Albany: New York State Department of Health.

Manual of Water Utility Operations, 8th ed. 1988. Austin: Texas Water Utilities Association. Available: www.twua.org/publications.htm#wuo.

National Atmospheric Deposition Program Annual Summary. 2001. Available: http://nadp.sws.uiuc.edu.

Spellman, F.R. 2008. *Handbook of Water and Wastewater Treatment Plant Operations.* Boca Raton, Fla.: CRC Press.

TABLE 1-2 Common measurements of water use—US customary units

Term	Definition	How to Calculate	Units of Measure	Comments on Use
Volume	Amount of water that flows through a treatment plant	Read totalizer at beginning and end of period for which volume is to be determined, and subtract first reading from second.	Gallons, millions of gallons, cubic feet, acre-feet	One of the basic measurements of water use
Instantaneous flow rate	The flow rate at any instant	Read directly from circular flow chart or from indicator.	Gallons per minute (gpm), gallons per hour (gph), gallons per day (gpd), million gallons per day (mgd), cubic feet per second (ft^3/sec)	One of the basic measurements of water use
Daily flow	Amount of water passing through plant during a single day	Measure flow volume (gallons or liters) for a single day.	gpd, mgd	Basis for calculating average daily flow and chemical designs
Average daily flow	Average of daily flows for a specified time period	Find sum of all daily flows for the period and divide by number of daily flows used.	gpd, mgd	Basis for calculating most other measurements
Annual average daily flow	Average of average daily flows for a 12-month period	Average the average daily flows for all the days of the year, or divide total flow volume for the year by 365.	gpd, mgd	Used to predict need for system expansion

Table continued next page

TABLE 1-2 Common measurements of water use—US customary units (Continued)

Term	Definition	How to Calculate	Units of Measure	Comments on Use
Peak-day demand	Greatest volume per day flowing through the plant for any day of the year	Look at records of daily flow rates for the year to find the peak day.	gpd, mgd	Determines system operation (plant output, storage) during heaviest load period—ranges from 1.5 to 3.5 times the average daily flow rate
Peak-hour demand	Greatest volume per hour flowing through the plant for any hour in the year	Determine from the chart recordings showing the continuous changes in flow rate.	gph	Determines required capacity of distribution system piping—ranges from 2.0 to 7.0 times the average hourly demand for the year
Minimum-day demand	Least volume per day flowing through the plant for any day of the year	Look at records of daily flow rates for the year to find the minimum day.	gpd, mgd	Determines possible plant shutdown periods—ranges from 0.5 to 0.8 times the average daily flow rate

Table continued next page

TABLE 1-2 Common measurements of water use—US customary units (Continued)

Term	Definition	How to Calculate	Units of Measure	Comments on Use
Filtration rate	Rate at which water is flowing through a filter	Divide the flow rate per minute flowing through the filter by the surface area of the filter.	gpm/ft^2	Indicates the rate of flow through a filter, particularly to check that the design rate is not exceeded
Peak-month demand	Greatest volume passing through the plant during a calendar month	Look at records of daily flow volumes (determined from totalizer readings at the beginning and end of each month or from the sums of the daily flows for each month) to find peak month for the year.	Gallons, millions of gallons, billions of gallons	Determines possible plant storage needs—ranges from 1.1 to 1.5 times the average monthly flow volume for the year
Minimum-month demand	Least volume passing through the plant during a calendar month	Look at records of monthly flow volumes (determined as for peak-month demand) to find minimum month for the year.	Gallons, millions of gallons, billions of gallons	Helps you to judge best time of year for checking and repairing equipment—ranges from 0.75 to 0.90 times the average monthly flow volumes for a year
Minimum-hour demand	Least volume per hour flowing through the plant for any hour in the year	Determine from the chart recordings showing the continuous changes in flow rate.	gpm, gph	Ranges from 0.20 to 0.75 times the average hourly demand for a year

CHAPTER 2

Groundwater Sources

A significant amount of the water used in North America is groundwater. About 20 percent of all water used in the United States, including water used for agriculture, mining, and power, comes from groundwater. About 22 percent of that amount is used for public and domestic water supply. Groundwater is the principal source of potable water for about 14 percent of the country's population, with 98 percent of the self-supplied domestic water withdrawals coming from groundwater (USGS Circular 1344 2009).

Groundwater sources can be relatively simple to develop, and the water itself often requires little or no treatment before use. Small public water systems usually find groundwater to be the most economical source if suitable aquifers are available.

WATER WELL TERMINOLOGY

Wells are the most common means of accessing groundwater. This section describes the various parts of a well and defines some common terms associated with wells.

Parts of a Well

Although wells come in many varieties and brands with different components, some of those components are quite typical (Figure 2-1). At the surface, all wells should have a sanitary seal, which prevents contamination from entering the well casing. The seal is generally a metal plate with a rubberized gasket around its perimeter that fits snugly into the top of the well casing. It has openings into the well for the discharge pipe, the pump power cable, and an air vent to let air into the casing as the water level drops. A sanitary seal for a small well with a submersible pump is illustrated in Figure 2-2.

The well casing is simply a liner placed in the borehole to prevent the surrounding rock and soil from collapsing into the shaft and blocking it. The casing (generally made of steel or plastic pipe) may not be necessary where the well extends into solid rock that has little risk of collapsing into the well. Where the well extends into solid rock, the casing is generally extended from the ground surface through the soil or seated into the rock to prevent soil from falling into the shaft.

The space between the casing and borehole above the production zone (i.e., where groundwater is withdrawn from the aquifer) is filled with grout to help prevent water from traveling along the outside of the casing and to support the casing. Where a well penetrates several aquifers, the grout also keeps water from flowing between aquifers along the outside of the well casing.

A well screen (intake screen) may be placed in a formation or highly fractured rock aquifer at intervals along the well where water intake is desired. The well screen prevents rock and soil from entering the well while letting in water. Often in a well that is screened

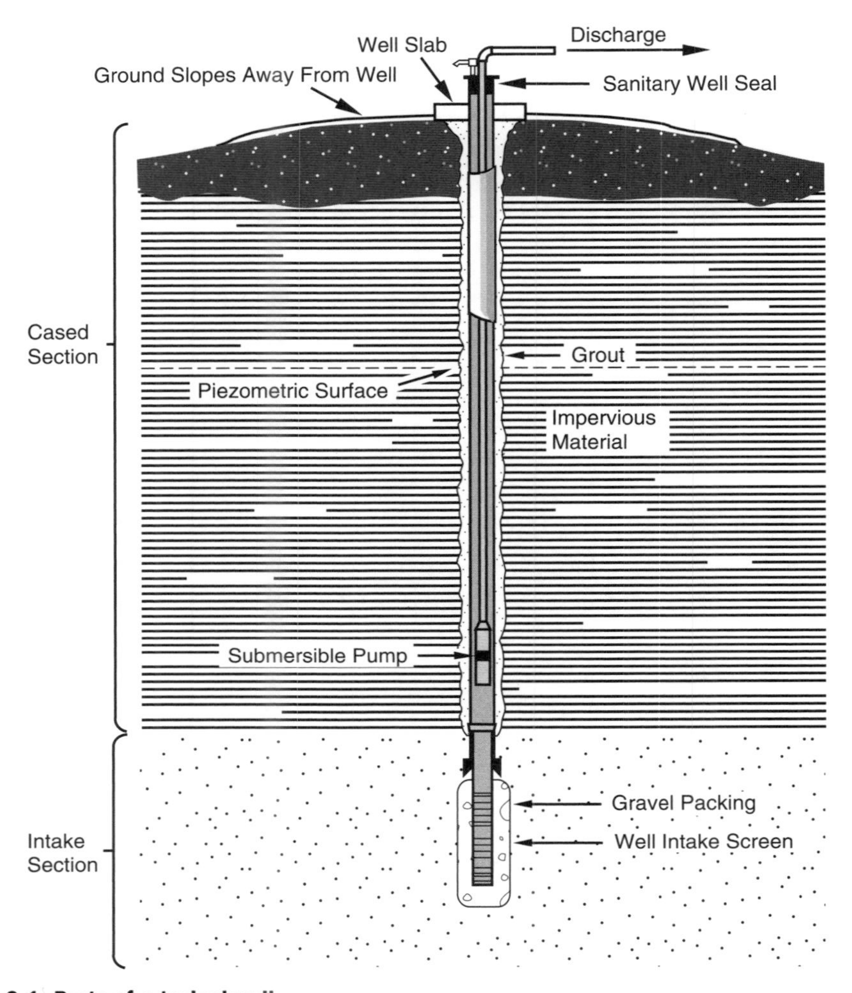

FIGURE 2-1 Parts of a typical well

in soil, sand is packed between the screen and the soil to fill the area and keep out sediment that would otherwise get into the well. Where the well has been drilled through solid rock, water may enter the well mainly through fractures in the rock rather than through the pores of the rock. In this case, the fractured zones may be screened to prevent rock chips around the fracture from falling into the well.

The well slab is a concrete area placed around the casing of some wells to support pumping equipment and to help prevent surface water from contaminating the well water.

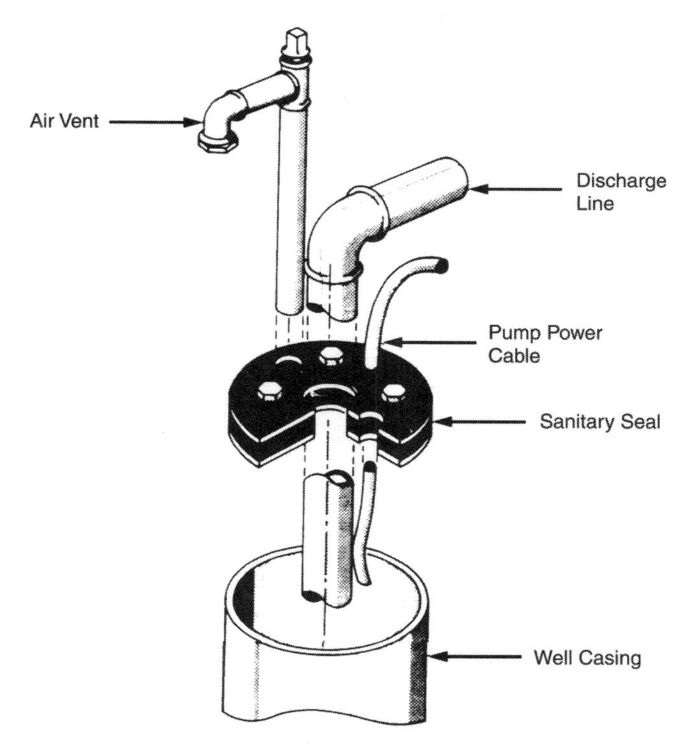

FIGURE 2-2 Components of a sanitary seal

Well Terms

Hydraulic characteristics of a well that are important during well operation are as follows:

- Static water level
- Pumping water level
- Drawdown
- Cone of depression
- Zone of influence
- Residual drawdown
- Well yield
- Specific capacity

These characteristics are discussed below. Several are illustrated in Figure 2-3.

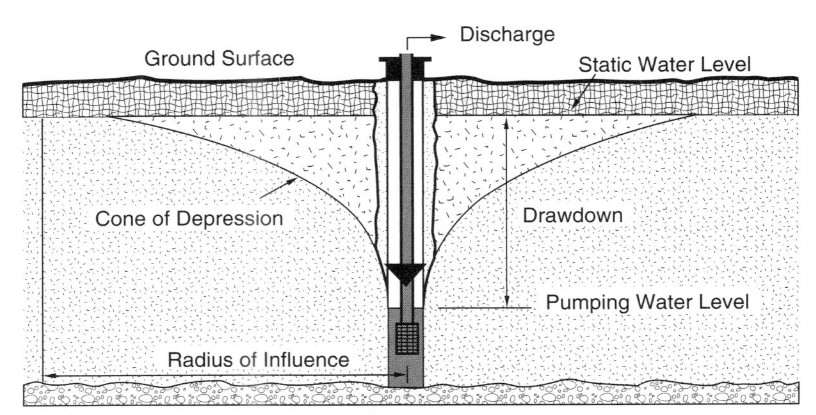

FIGURE 2-3 Hydraulic influence of a well on an aquifer

Static water level

The static water level in a well is the level of the water surface in the well when no water is being taken from the aquifer. It is normally measured as the depth from the ground surface to the water surface and may be converted to elevation for comparison with information about other wells in the area. This is an important measurement because it is the basis for monitoring changes in the water table.

In some areas, the water table is just below the ground surface; usually, it is several feet below the surface. In other areas, there is no groundwater near the surface, and a well may not pass through any water until an aquifer is intercepted at hundreds or even thousands of feet below the surface.

Pumping water level

When water is pumped out of a well, the water level usually drops below the level in the surrounding aquifer and eventually stabilizes at a lower level called the pumping level. The level varies depending on the pumping rate. The water intake or submerged pump must be located below this level. It is also preferred to limit the pumping rate so that this level is above the screened portion of the well (or, in an open rock well, above water-bearing fractures) to prevent water from continually cascading down the sides of the well.

Drawdown

The drop in water level between the static water level and the pumping water level is called the drawdown of the well.

Cone of depression

In unconfined aquifers, water flows from all directions toward the well during pumping. The free water surface in the aquifer then takes the shape of an inverted cone or curved funnel called the cone of depression. In a confined aquifer, the water pressure within the aquifer will also be much less at the well than at some distance from the well. A graph of the pressure versus distance from the well will also have the shape of a curved funnel that is centered on the well.

Zone of influence

The size of the area that is affected by drawdown depends on the porosity, conductivity, and other properties of the aquifer. When drawdown occurs, the water table (or pressure, in the case of an unconfined aquifer) is affected significantly out to some distance away from the well. That horizontal distance is called the radius of influence (Figure 2-3), and the entire affected area is called the zone of influence.

If the aquifer is composed of material that transmits water easily, such as coarse sand or gravel, the cone of depression may be almost flat and the zone of influence relatively small. If the material in the aquifer transmits water slowly, the cone will usually be quite steep and the zone of influence large.

If possible, wells should be situated far enough apart that their zones of influence do not overlap. The effect on the cones of depression of two wells located close together is illustrated in Figure 2-4. Pumping a single well reduces the water level in the other well, and pumping both simultaneously creates an undesirable interference in the drawdown profiles.

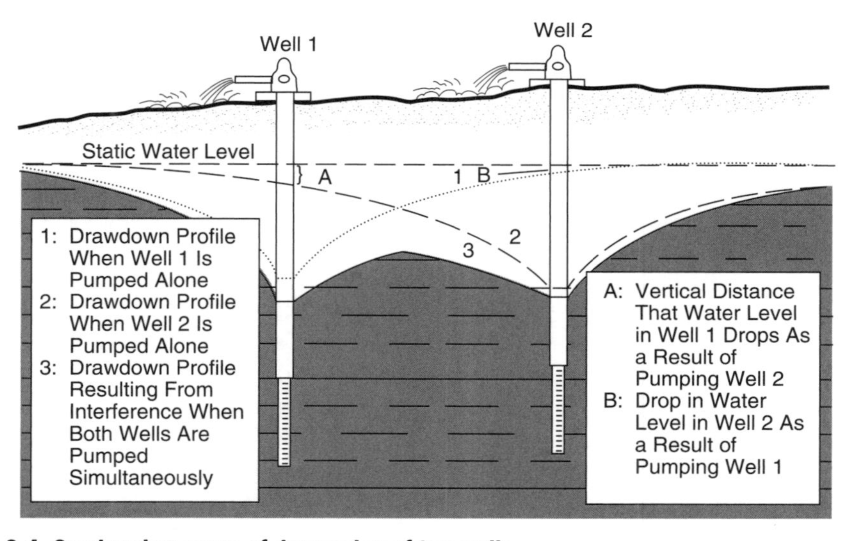

FIGURE 2-4 Overlapping cones of depression of two wells

Residual drawdown

After the pumping of a well has stopped, the water level in the well rises back up toward the static water level. If the water level does not quite reach the original level, the distance it falls short is called the residual drawdown.

Well yield

Well yield is the rate of water withdrawal that a well can supply over a long period of time. The yield of small wells is usually measured in gallons per minute (liters per minute) or gallons per hour (liters per hour). For large wells, it may be measured in cubic feet per second (cubic meters per second).

When more water is taken from an aquifer than is replaced by recharge, the drawdown gradually reaches greater depths and the safe well yield is reduced. Under prolonged pumping, the drawdown can get so low that the pump begins to suck air, which will damage it. To prevent this damage to the pump, water can be kept at a safe level by limiting the pump rate, lowering the pump depth, or operating the pump for shorter periods of time.

Ideally, wells should pump continuously with no permanent drawdowns. A more common practice is to pump wells that have a significant drawdown for only a few hours each day, allowing an extended period for the aquifer to recover. But even this is difficult, as the long-term capacity of an aquifer can be determined by short-term tests and study of the geology of the area. Rarely is a consultant paid to determine that sustainable water supplies are not available. The result is the potential for aquifer drawdown accompanied by aquifer mining and land subsidence. Confounding the problem is that many aquifer systems cross political boundaries, so careful regulation in one jurisdiction may not be supported by others.

Reilly et al, (2009) is USGS Report 1323 published to outline the current condition of groundwater in the United States. Reilly et al, 2009 estimates that the pumpage of fresh ground water in the United States is approximately 83 billion gallons per day (Hutson and others 2004), which is about 8 percent of the estimated 1 trillion gallons per day of natural recharge to the Nation's ground-water systems (Nace 1960), which sounds like it is not a serious issue. However, Reilly et al, 2009 found that the loss of groundwater supplies in many areas will be catastrophic, affecting economic viability of communities and potentially disrupting lives and ecological viability. Figure 2-5 combines regional water-level declines and local water-level declines for changes on a national scale and shows water-level declines over the last 40 years throughout the United States. The Great Plains states, Texas and the west are particularly affected. The red regions indicate areas in excess of 500 square miles that have water-level decline in excess of 40 feet in at least one confined aquifer since predevelopment, or in excess of 25 feet of decline in unconfined aquifers since predevelopment. Blue dots are wells in the USGS National Water Information System database where the measured water-level difference over time is equal to or greater than 40 feet. This would indicate that groundwater in these area are not sustainable and therefore will become less available with time. Given the critical nature of this issue, Reilly

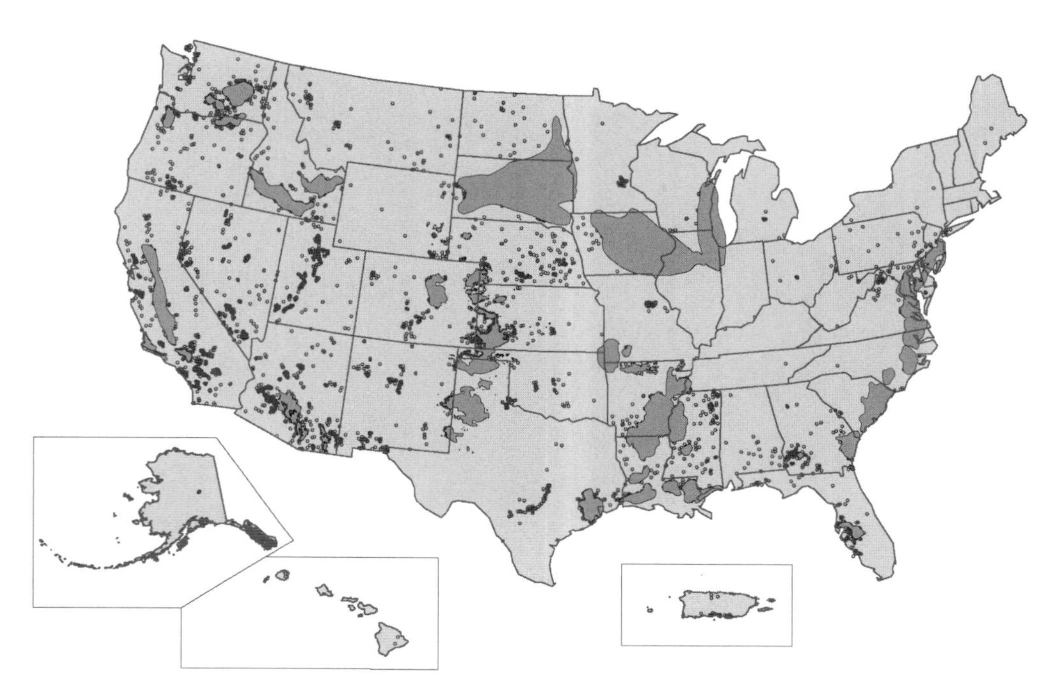

FIGURE 2-5 Regional and local water-level declines on a national scale

et al, (2009) advised that a coordinated nation-wide effort to organize available information on changes in groundwater storage is needed..

Bloetscher (2009) notes that drilling deeper is not a solution. Deeper waters tend to have poorer water quality, having been in contact with the rock formation longer and absorbed more dissolved minerals. Therefore, while a deep aquifer may be prolific, water obtained from a well that taps into it may not be desirable or potable without substantial amounts of treatment. In addition, most deeper aquifers are confined and therefore do not recharge significantly locally. The withdrawal of water may appear to be a permanent loss of the resource in the long term. For example, portions of the Black Creek aquifer in eastern North and South Carolina were virtually depleted by pumpage because there is no local recharge. As a result the aquifer was mined, exceeding its safe yield, and the large utilities converted to surface water. Likewise, most of the aquifers in the western United States have minimal potential for recharge. In parts of the western plains states and Great Basin, the aquifers have dropped hundreds of feet (meters). With an average of 13–18 inches (33–46 cm) per year of rainfall and high evaporation rates throughout the summer, there is little potential that the aquifer will be recharged (Bloetscher and Muniz 2008).

The current evaluation suggests the likelihood of conflicts over water supplies in the near future. To reduce this potential, laws must be enacted that resolve water rights and water quality issues. In the absence of negotiated settlements between the parties in conflict over water rights or withdrawal rights, these cases often go to court and are not likely to be quickly decided there.

Identifying critical natural capital requires a systematic analysis and evaluation of whether environmental resources are being used sustainably, the extent of any sustainability gaps, the economic and environmental pressures, and public policies aimed at improving ecological systems. Ekins (2003) suggests criteria that can be used:

- Maintenance of human health to avoid negative health impacts
- Avoidance of loss of ecological function
- Economic sustainability—maintenance of economic activities on basis that does not deplete the resource

Specific capacity

One of the most important concepts in well operation and testing is specific capacity, which is a measure of well yield per unit of drawdown. It can be calculated as follows:

$$\text{specific capacity} = \text{well yield} \div \text{drawdown}$$

For example, if the well yield is 200 gpm (757 L/min) and the drawdown is measured to be 25 ft (7.6 m), the specific capacity is

$$200 \div 25\ (757 \div 7.6) = \text{specific capacity}$$

$$\text{specific capacity} = 8\ \text{gpm}\ (30\ \text{L/min})\ \text{per 1 ft}\ (0.3\ \text{m})\ \text{of drawdown}$$

The calculation is simple and should be made frequently in the monitoring of well operation. A sudden drop in specific capacity indicates trouble such as pump wear, screen plugging, or other problems that can be serious. These problems should be identified and corrected as soon as possible.

WELL LOCATION

A well should be located to produce the maximum yield possible without introducing any kind of contamination. Several methods can be used to determine likely locations for a new well.

Existing Data

When a new well is to be constructed, among the first places to look for helpful information about aquifers in the area are the state and federal geological or water resources agencies. In most states, data from previous well logs have been used to draw maps of the geology of various regions and to identify water-bearing layers. The agencies should also be of assistance in establishing the productivity of aquifers in the region and determining which of them are capable of providing adequate flow to additional wells.

These agencies will probably also have information on water quality, including data about hardness and about concentrations of iron and manganese, sulfur, nitrate, and radionuclides, as well as information about other water characteristics that might be of concern. The state agency may also be able to provide some information on the possibility of contamination of the aquifer from industrial or agricultural chemicals.

The owners of public and private wells that already exist nearby may also provide useful information. If there is more than one aquifer in the area, it is important that the information obtained applies only to wells that will draw from the aquifer being considered for the new well. Although there are often some differences in water quality within an aquifer, the information from existing wells already drilled into that aquifer will provide a good basis for knowing what to expect from a new well.

Local well-drilling contractors are the third source of information that should be consulted. They usually have extensive practical information about where wells can be developed and the quantity and quality of water that will be obtained.

A licensed hydrogeologist can provide invaluable professional help in determining the most likely location to drill a well that suits the needs of the owner. This is especially true in areas where water is scarce or no large aquifers exist. A local hydrogeologist should be familiar with the area's geologic and groundwater features and should have some knowledge about regulations regarding the permitting of water supply wells.

Likely Locations

As one might expect, groundwater is likely to be present in larger quantities beneath valleys rather than under hills. Valley soils containing permeable material washed down from mountains are usually productive aquifers. In some areas of the country, the only groundwater of usable quality is found in river valleys. Also, the bedrock underlying valleys is often fractured and water bearing. Coastal terraces and coastal and river plains may also have good aquifers.

Any evidence of surface water, such as streams, springs, seeps, swamps, or lakes, is a good indicator that groundwater is present, though not necessarily in usable quantities. Sometimes the presence of vegetation will reveal the location of groundwater; an unusually thick overgrowth may indicate shallow groundwater.

Exploration

After all existing data and likely locations have been examined, there still may not be enough information to risk spending a large amount of money on a well. In that case, it may be worthwhile to conduct an underground exploration to provide additional information about local geologic formations, and aquifers in particular. Some of the more common methods include geophysical exploration, test wells, and computer modeling.

Geophysical exploration

One approach to getting more information about groundwater in a given locale is to have a geophysical survey performed professionally. The science of geophysics uses indirect means to characterize the subsurface and is often a cost-effective way to reveal the location and extent of geologic formations and features. Understanding the overall geologic picture often improves understanding of where the best location to drill a well might be. Features that suggest a good place for a well include water-bearing faults and fracture zones in the bedrock and the lateral and vertical extent of specific aquifers.

A professional geophysicist, in coordination with a hydrogeologist, can determine the most appropriate geophysical survey method—seismic or resistivity—for the specified purpose and location. Seismic methods measure the speed at which a shock wave travels through the earth. A small charge of dynamite is discharged at the bottom of a drilled shot hole. Instruments record the time it takes for the shock waves to pass through the ground and reach vibration sensors (called geophones) placed at spaced distances at the surface (Figure 2-6). The time recorded at the various locations can be used to determine where water-bearing formations might be located. Resistivity methods measure the ground's electrical resistance with depth into the ground. In general, the lower the electrical resistance, the greater the probability that water is present. Both tests can be conducted on the ground surface, without the need to drill test holes.

Test wells

If it appears likely that water is present, exploration holes can be drilled in key locations. Samples should be kept of the material removed as drilling progresses on each hole. The holes may then be logged (i.e., tested) electrically or by using gamma rays to help define water-bearing formations. Electric logging measures the change in the earth's resistance to an electrical current as the depth increases. Gamma-ray logging measures how controlled radiation penetrates the earth. Information from the logs is correlated with the samples removed during the drilling to identify the various formations below the surface.

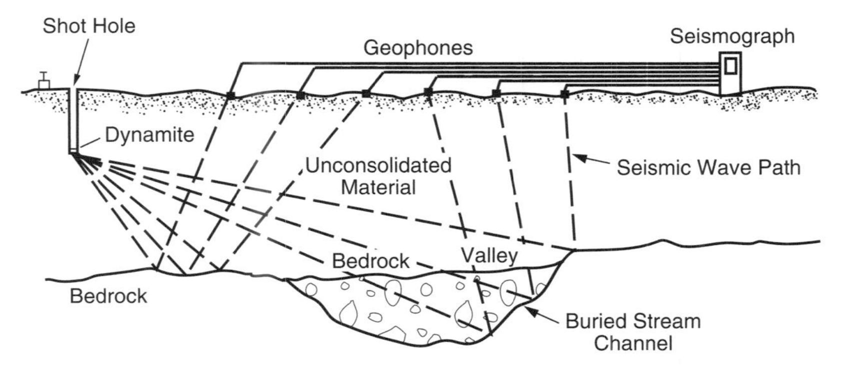

FIGURE 2-6 Application of the seismic refraction method for reconnaissance mapping

Computer modeling

Hydrologists interested in evaluating the many complex withdrawal stresses and effects on an aquifer system can use computer models to perform studies that were not previously possible with manual calculations. This type of analysis might be necessary, for example, to determine an allowable maximum yield in areas where several wells are stressing the aquifer. Computer analyses using available information on aquifer recharge and withdrawals can also help to determine the most productive locations for additional wells, forecast the effects of contamination sources on the aquifer, and identify other factors that will affect the success of a well-drilling effort. Computer models can be expensive to develop independently, however. They are most widely used by government agencies to develop area and regional plans for groundwater use (Figure 2-7).

Preventing Well Contamination

Wells should be located, built, and operated in a way that prevents contamination of the well water. Contamination may be biological, as in the introduction of disease-causing organisms, or chemical, as in the introduction of toxic materials. The consequences of contamination can vary considerably, from mild changes in taste and odor to severe illness. Generally, federal and state environmental agencies establish maximum contaminant levels (MCLs) for various uses of water.

New wells must be located a safe distance from sources of potential contamination. Contamination can be the result of human activities near or at the surface. An underground fuel storage tank, for example, will contaminate the ground and groundwater if it leaks. However, contamination may also be the result of overpumping, as in the case of salt-water intrusion or surface water capture. Naturally occurring substances such as radon, arsenic, or iron can make well water unfit for use if they occur in concentrations high enough to pose a risk.

Typically, to prevent contamination, a minimum radial distance from the top of the well (wellhead protection area) is regulated by local and state environmental agencies. The regulations generally limit the type and extent of activities that can be conducted within the wellhead protection area.

Existing wells sometimes become contaminated because of new contamination sources or sources that were not anticipated when the well was installed. Potential sources of groundwater contamination are listed in Table 2-1.

As noted, many of these contaminants may be naturally occurring or synthetic and may be located on or below the ground surface. Contamination could be introduced out of ignorance, by accident, or by intentional illegal disposal. Contaminants can also be carried into an aquifer with percolating surface water and can travel considerable distances.

Because of differences in soil permeability, the depth of soil layers, and types of underlying rocky materials, as well as the varying rates and directions of groundwater movement, there is no set safe, protective distance that applies to all wells. Some chemical contaminants have been tracked more than 0.5 mi (0.8 km) from their source.

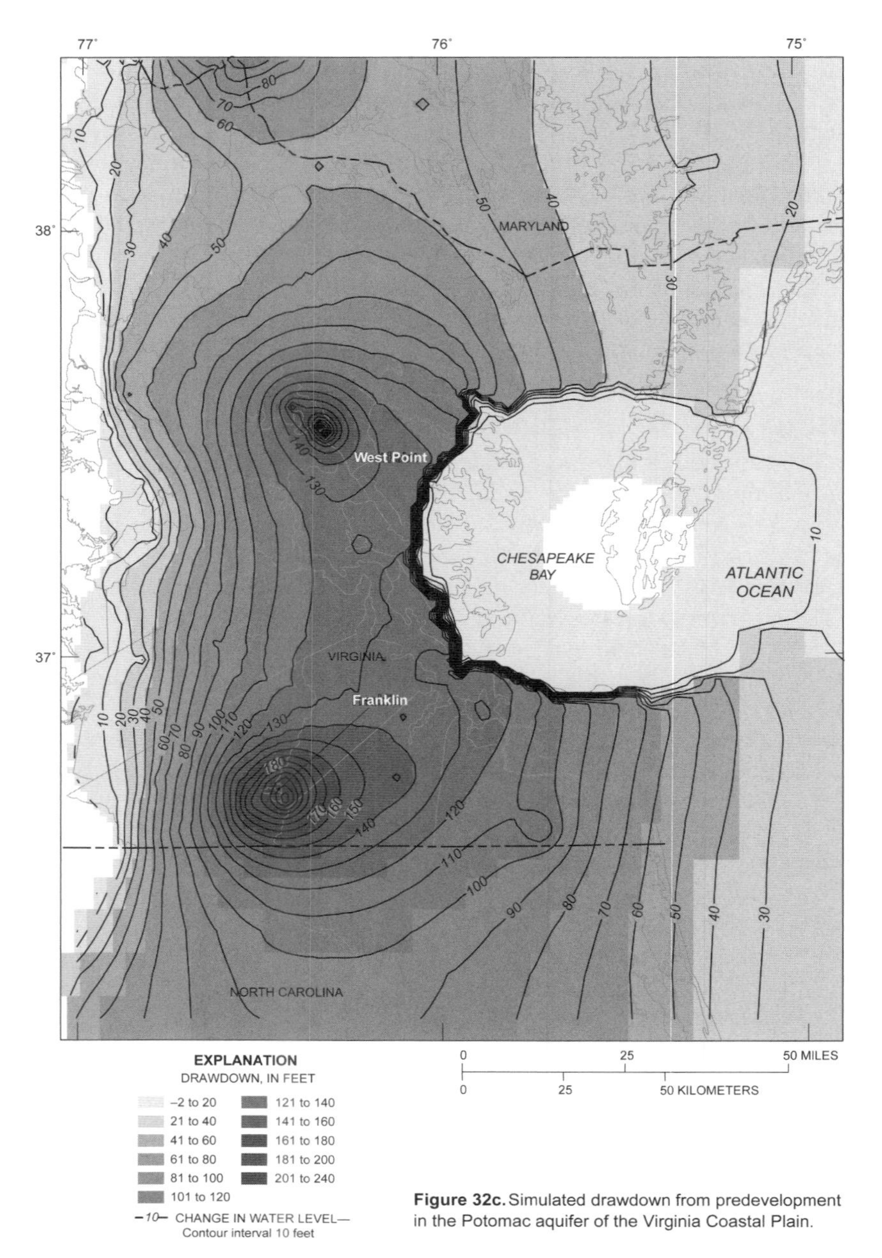

FIGURE 2-7 Computer modeled drawdown from predevelopment to 2003 in the Potomac Aquifer of the Virginia Coastal Plain

TABLE 2-1 Potential sources of groundwater contamination

Source	Possible Major Contaminants
Landfills	
Municipal	Heavy metals, chloride, sodium, calcium
Industrial	Wide variety of organic and inorganic constituents
Hazardous-waste disposal sites	Wide variety of inorganic constituents (particularly heavy metals such as hexavalent chromium) and organic compounds (pesticides, solvents, polychlorinated biphenyls)
Liquid waste storage ponds (lagoons, leaching ponds, and evaporation basins)	Heavy metals, solvents, and brines
Septic tanks and leach fields	Organic compounds (solvents), nitrates, sulfates, sodium, and microbiological contaminants
Deep-well waste injection	Variety of organic and inorganic compounds
Agricultural activities	Nitrates, herbicides, and pesticides
Land application of wastewater and sludges	Heavy metals, organic compounds, inorganic compounds, and microbiological contaminants
Infiltration of urban runoff	Inorganic compounds, heavy metals, and petroleum products, along with microbiological contaminants
Deicing activities (control of snow and ice on roads)	Chlorides, sodium, and calcium
Radioactive wastes	Radioactivity from strontium, tritium, and other radionuclides
Improperly abandoned wells and exploration holes	Variety of organic, inorganic, and microbiological contaminants from surface runoff and other contaminated aquifers

Source: Basics of Well Construction Operator's Guide

When siting a new well, one must keep in mind that when the well is put into operation, the direction of groundwater flow could change. As illustrated in Figure 2-8, water flow in an aquifer that appears to be coming from a direction with no contamination can actually be reversed by a new well's zone of influence. This illustrates the need for continual sampling and analysis to check for potential contamination of public supply water sources.

Site Selection

Ultimately, the selection of a well site depends not only on the geology of the region and the need to prevent contaminants, but also on legal and political considerations. In a residential community, for instance, the decision of where to site a new well must also be based on where land can be acquired and where the construction of a well will be aesthetically acceptable to residents.

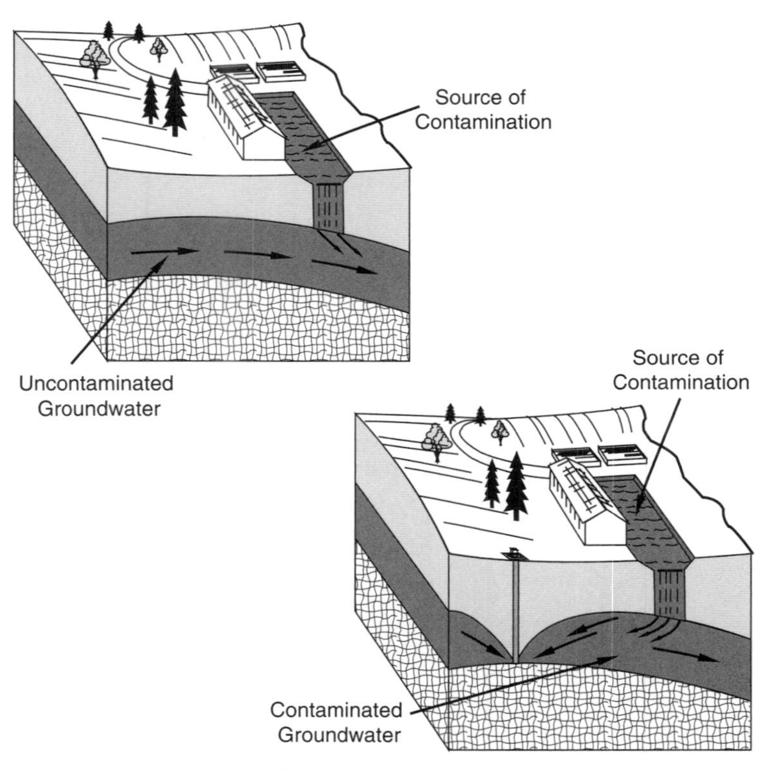

FIGURE 2-8 Reversal of flow in an aquifer due to well drawdown

TYPES OF WELLS

Wells are generally classified according to their type of construction as follows:

- Dug wells
- Bored wells
- Driven wells
- Jetted wells
- Drilled wells

Dug Wells

A dug well can furnish large quantities of water from shallow groundwater sources. Dug wells are uncommon because of the cost to install them and the potential for contamination of the shallow aquifers and the wells themselves. Small-diameter wells can be constructed manually with pick and shovel; larger wells are constructed with machinery such as a clamshell bucket if the soil conditions are suitable. Figure 2-9 illustrates a typical dug-well installation.

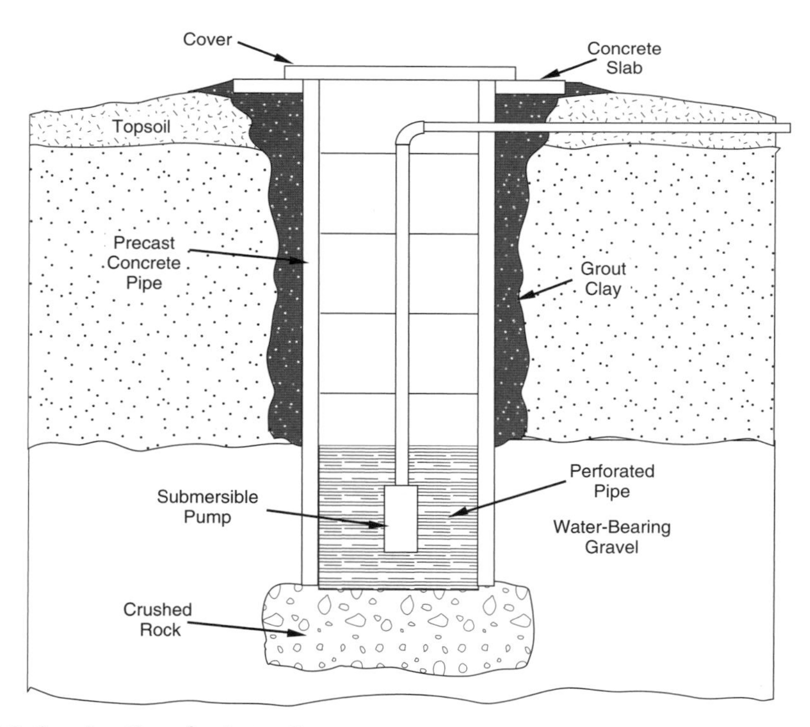

FIGURE 2-9 Construction of a dug well

If the exposed soil will stand without support, it may not be necessary to line the excavation until the water table is reached. Precast or cast-in-place concrete liners, commonly called curbs, are used to seal the shaft. The liners in contact with the water-bearing layers are perforated to allow water to enter. If the aquifer is sandy material, a layer of gravel is placed around the curb to act as a sand barrier.

Yield from a dug well increases with an increase in diameter, but the increase is not directly proportional. Dug wells serving a public water system may be 8 to 30 ft (2 to 9 m) in diameter and from 20 to 40 ft (6 to 12 m) deep.

Most dug wells do not penetrate much below the water table because of the difficulty of excavating in the saturated soil. For this reason, a dug well may fail if the water level recedes during times of drought or if there is unusually heavy pumpage from the well.

The surface opening of a dug well is large, making protection from surface contamination difficult. For that reason, state regulatory agencies usually classify these wells as being vulnerable to contamination, and the wells must be treated as a surface water source. Disinfection and possibly filtration treatment will be required if the water source is to be used by a public water system.

Bored Wells

Wells can be constructed quickly by boring where the geologic formation types are suitable. The formation must be soft enough for an auger to penetrate yet firm enough so it will not cave in before a liner can be installed. The most suitable formations for bored wells are glacial till and alluvial valley deposits.

A bored well is constructed by driving an auger into the earth. Bored wells are limited to approximately 3 ft (1 m) in diameter and depths of 25 to 60 ft (8 to 18 m) under suitable conditions. As the auger penetrates, extensions are added to the drive shaft. A casing is forced into the hole as material is removed until the water-bearing strata are reached. Installing well screens or a perforated casing in the water-bearing sand and gravel layer completes the well.

Cement grout is used to surround the casing to prevent entrance of surface water, which could cause contamination. Bored wells are infrequently used for public water supplies.

Driven Wells

Driven wells are simple to install; however, they are practical only when the water-bearing formations are relatively close to the surface and no boulders or bedrock exist in the soil or other formations between the surface and the aquifer. These wells consist of a pointed well screen, called a drive point, and lengths of pipe attached to the point. The point has a steel tip that enables it to be pounded through some gravel or hardpan (a hard layer of cemented soil near the ground surface) to the water-bearing formation (Figure 2-10).

The diameter of the well pipe varies from as small as 1 in. (32 mm) up to 4 in. (107 mm). The maximum depth that can be achieved is generally 30 to 40 ft (9 to 12 m). This well type

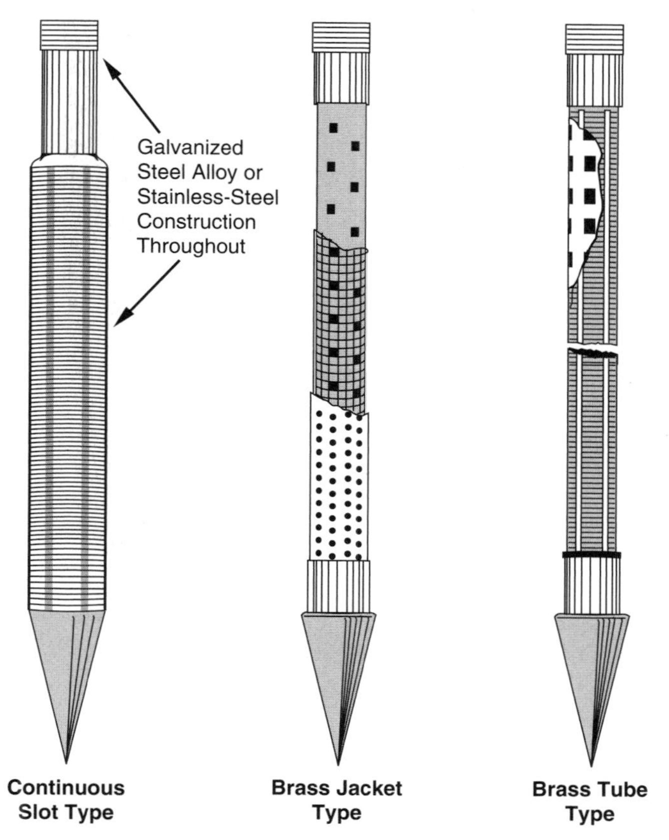

FIGURE 2-10 Different types of driven-well points
Source: Manual of Individual Water Supply Systems (1982).

is not generally used for public water supply because of the potential for environmental contamination of shallow aquifers and because of their relatively small diameter—a single well cannot produce the quantity of water generally needed for public supply. However, a battery of points may be used for greater production, with several wells connected by a common header to a suction-type pump. A suction pump can be used only when the static water level is no deeper than about 15 ft (5 m).

A smaller-diameter pilot hole is usually drilled or driven before driving the well to make installation easier and help prevent damage to the well casing. This casing of durable pipe can be pounded into the earth or placed in a bored hole. The casing prevents water contamination if there should be leaky joints in the well pipe.

After the outer casing is in place, the inner casing with a perforated well point is inserted. The drive point is then driven into the water-bearing formation.

Jetted Wells

A jetting pipe, which is equipped with a cutting knife on the bottom, is used to construct jetted wells. Water is pumped down the pipe and out of the drill bit against the bottom of the hole. The high-pressure water jet at the bottom of the pipe, in coordination with the cutting knife, loosens and removes the soil beneath the pipe and allows it to advance downward.

The casing is usually sunk as the drilling progresses until it passes through the water-bearing formation. The well screen connected to a smaller-diameter pipe is then lowered into the casing, and the outer casing is withdrawn to expose the screen to the formation. These wells are generally suited to sandy formations and cannot be constructed by jetting through clay or hardpan or where there are boulders in the formation.

Drilled Wells

Drilled wells are the most commonly used well type for public water supplies because they can be installed in almost any situation. They can be constructed to extreme depths with small or large well diameters (up to 4 ft [1.5 m] or possibly larger). They are also the most common type of well drilled for oil extraction.

A drilled well is constructed using a drilling rig and casing. The rig makes the hole, and casing is placed in the hole to prevent the walls from collapsing. Screens are installed when water-bearing formations are encountered at one or more levels. The more commonly used methods of drilling water supply wells are the following:

- Cable tool method
- Rotary hydraulic method
- Reverse-circulation rotary method
- California method
- Rotary air method
- Down-the-hole hammer method

Cable tool method

The percussion drilling method, commonly referred to as the cable tool method, has been used extensively for wells of all sizes and depths but has waned in popularity because faster and easier methods have been developed. There are many commercial varieties of cable tool rigs. The operating principle for all varieties is the same. They use a bit at the end of a cable that is repeatedly raised and dropped to fracture the soil material. The drilling tool has a clublike chisel edge that breaks the formation into small fragments. The reciprocating motion of the drilling tool mixes the loosened material into a sludgelike substance.

In each run of the drill, a depth of 3–6 ft (1–2 m) of the hole is drilled. The drill is then pulled from the hole, and a bailer is used to remove the sludge. The bailer consists of a section of casing 10 to 25 ft (3 to 8 m) long, slightly smaller in diameter than the drilled hole, and having a check valve in the bottom. A casing is forced into the hole as soon as it is necessary to prevent a cave-in of the walls.

As the drill operates, an operator continually adjusts the length of stroke and rapidity of blows based on experience and the feel of the vibrations. A skilled operator will be able to distinguish the hardness of the formation being drilled by the vibrations in the cable and can sense the passing of the drill from one formation to another. As drilling progresses, samples of the cuttings will be taken periodically to check on the type of formation being penetrated.

After drilling the well to the maximum desired depth, a screen is lowered inside the casing and held in place while the casing is pulled back to expose the screen (Figure 2-11).

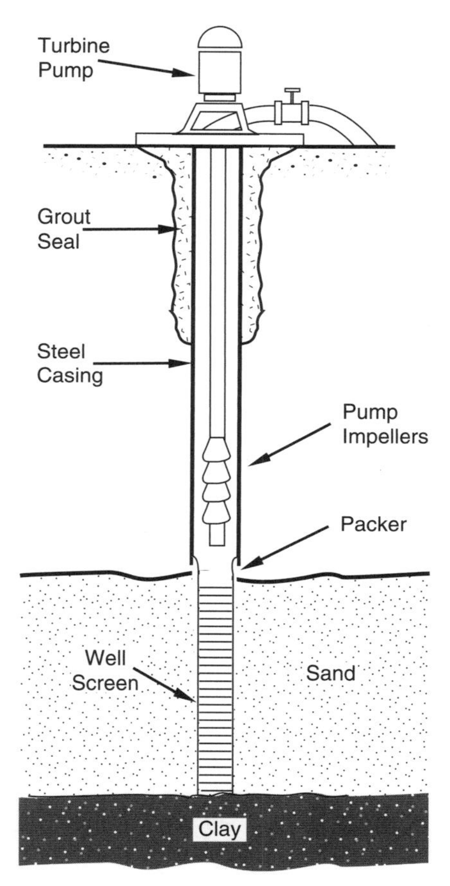

FIGURE 2-11 Exposed well screen

The top of the screen is then sealed against the casing by expanding a packer of neoprene rubber or lead. When a well reaches consolidated rock, the normal practice is to seat the casing firmly in the top of the rock (providing a secure seal where the casing meets the rock) and drill an open hole to the depth required to obtain the needed yield (Figure 2-12).

Rotary hydraulic method

In the rotary method of well drilling, the hole is made by spinning a cylinder-shaped bit on the bottom of multiple sections of drill pipe. The speed of rotation can be varied to achieve the best cutting effectiveness for different types of soil and rock.

Drilling fluid (typically a thin slurry of clay and water, also called drilling mud) is pumped down to the bit (Figure 2-13). The fluid flows out through holes in the bit, picks up loosened material, and carries it up the borehole to the surface. The circulating fluid also helps to cool the bit and keep the hole open during drilling. The fluid that flows to the surface overflows from the well and is routed by a ditch into a settling pit or tub, where the cuttings settle out. The fluid can then be reused.

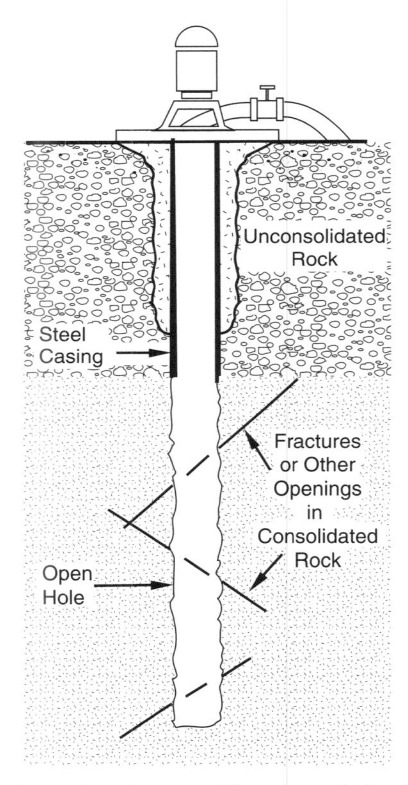

FIGURE 2-12 Casing seated at top of rock layer with an open hole underneath

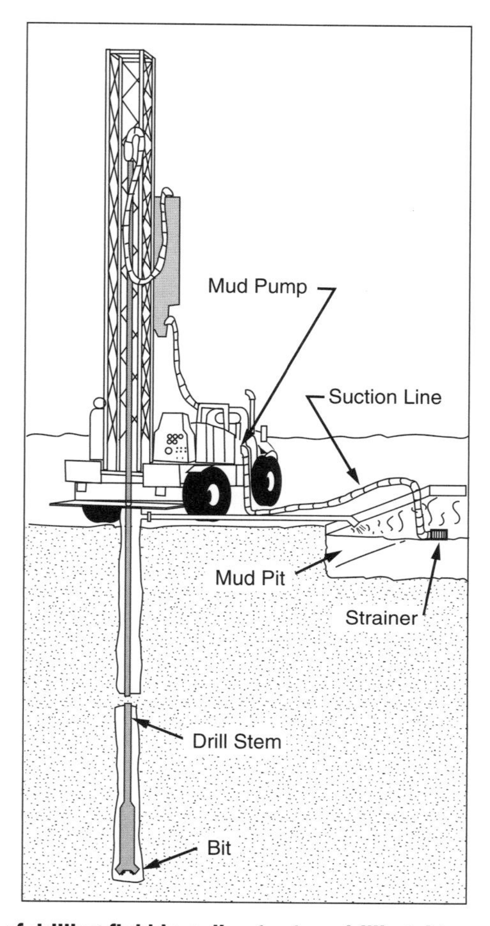

FIGURE 2-13 Circulation of drilling fluid in a direct rotary drilling rig

Clay that is added to the drilling fluid will adhere to the sides of the hole and, together with the pressure exerted by the drilling fluid, prevents a cave-in of the formation material. In some cases, native clay can be used; otherwise, prepared materials such as bentonite (a commercially available, highly expansive clay) or a special powdered substance is used. The drilling fluid is prepared in a pit near the drilling operation before it is pumped into the well. Its composition can be varied depending on the material that is being drilled through.

Reverse-circulation rotary method

The reverse-circulation rotary method of drilling differs from the regular rotary method in that the drilling fluid is circulated in the opposite direction. The advantage is that because water is generally used as the drilling fluid, the method can be used where drilling additives are undesirable. This method is also particularly well suited for constructing gravel-packed wells, because the drilling fluid mixture used in the regular rotary method tends to plug the walls of the well.

Drilling fluid is forced down through the borehole and returns to the surface in the hollow drill pipe, carrying the cuttings with it. A pump with a capacity of 500 gpm (1,900 L/min) or more is required to keep the fluid moving at high velocity. If an abundant freshwater supply is available, the discharge can be diverted to waste and not recycled. Otherwise, the cuttings are allowed to settle in a large pit and the fluid is recirculated.

Both rotary methods can drill holes up to 5 ft (1.5 m) in diameter, and the drilling is usually faster than cable tool drilling.

California method

The California method, also called the stovepipe method, was developed primarily for sinking wells in unconsolidated material such as alternating strata of clay, sand, and gravel. The process is similar to the cable tool method except that a special bucket is used as both bit and bailer. Each time the bit is dropped, some of the cuttings are trapped in the bailer. When the bailer is filled, it is raised to the surface and emptied.

The process also uses short lengths of sheet metal for casing. The casing is forced down by hydraulic jacks or driven by means of cable tools. After the casing is in place, it is perforated in place using special tools.

Rotary air method

The rotary air method is similar to the rotary hydraulic method except that the drilling fluid is air rather than a mixture of water and clay. The method is suitable only for drilling in consolidated rock. Most large drill rigs are equipped for both air and hydraulic drilling so that the method may be changed as varying strata are encountered.

Down-the-hole hammer method

A method frequently chosen to drill wells into rock uses a pneumatic hammer unit that is attached to the end of the drill pipe. The hammer is operated by compressed air. The air also cleans the cuttings away from the bit and carries them to the surface. For most types of rock, this is the fastest drilling method available.

The drilling rig for this method must be furnished with a very large air compressor. Most standard rigs use 750 to 1,050 ft^3/min (0.35 to 0.49 m^3/sec) of air at a pressure of 250 psi (1,700 kPa). Some rigs are also capable of operating at 350 psi (2,400 kPa), which will advance the drill twice as fast as a standard rig. A drilling rig of this type is illustrated in Figure 2-14.

FIGURE 2-14 A truck mounted drilling rig
Courtesy of Ingersoll-Rand Company

Special Types of Wells

The names of several types of wells refer not to their method of construction, but to other characteristics. Two such types are discussed here:

- Radial wells
- Bedrock wells

Radial wells

Radial wells are commonly used near the shore of a lake or near a river to obtain a large amount of relatively good-quality water from adjacent sand or gravel beds. Radial wells are also used in place of multiple vertical wells to obtain water from a relatively shallow aquifer or to obtain water from a surface body in such a way that the sediment underneath the river or lake acts as a filter. A radial well can be described as a dug well that has horizontal wells projecting outward from the bottom of the vertical central well (Figure 2-15). The central well, or caisson, serves as the water collector for the water produced by the horizontal screened wells.

Construction of a radial well begins by sinking the central caisson, which is generally 15 to 20 ft (5 to 6 m) in diameter. The caisson is made by stacking poured-in-place reinforced concrete rings, each about 8 to 10 ft (2 to 3 m) high. The first section is formed with a cutting edge to facilitate the caisson's settling within the excavation. As sediment is excavated from within the caisson, the concrete ring sinks into place. Additional sections are then added and the excavation progresses. When the desired depth is reached, a concrete plug is poured to make a floor.

Horizontal wells are then constructed through wall sleeves near the bottom of the water-bearing strata. The laterals may be constructed of slotted or perforated pipe, or they may have conventional well screens. Each horizontal well is constructed with a gate valve located inside the caisson to enable subsequent dewatering or to be closed if use of individual collector wells is discontinued. A superstructure is then erected on top of the caisson to house pumps, piping, and controls.

Bedrock wells

Bedrock wells are drilled into the underlying bedrock. Water flows to the well through fractures in the bedrock rather than from a saturated layer.

WELL CONSTRUCTION PROCEDURES

Wells can be constructed using several procedures. The principal factors affecting the choice of construction method are how deep the well must be and whether the material around the well is gravel, clay, or rock. After a well has been constructed, it is usually necessary to specially treat the well to obtain optimum productivity and water quality. The last step in well construction is to run pumping tests to confirm the well's capacity.

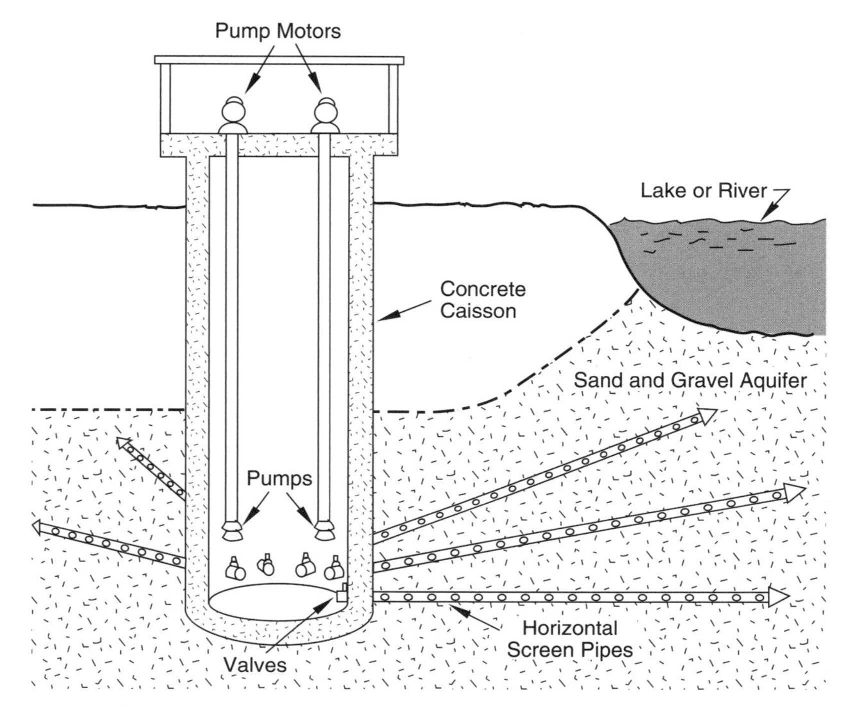

FIGURE 2-15 Details of a radial well

Well Components

Components common to most wells are:

- Well casings
- Gravel pack
- Well screens
- Grouting

Well casings

A well casing is a lining for the drilled hole that maintains the opening from the land surface to the water-bearing formation. The casing also helps prevent surface water from contaminating the water being drawn from the well.

Materials commonly used for well casings include alloyed or unalloyed steel, fiberglass, and plastic. The principal factors determining the suitability of casing materials are the stress that will be placed on the casing during installation and the corrosiveness of the water and soil that will be in contact with the casing.

When a well is being drilled by the cable tool method, the casing is driven as soon as it becomes necessary to prevent the walls of the well from caving in. A drive shoe of hardened steel is attached to the lower end of the pipe, and a drive head is attached to the top of the pipe to withstand the hard blows of driving the casing into the ground. As the drilling progresses, the casing is continually pounded by the action of the drilling equipment.

Wells constructed using rotary methods are not usually cased until the well hole has been completed. A casing diameter smaller than the hole is used so that the casing does not have to be driven.

If additional protection from corrosion and surface water pollution is required, an outer casing is first installed and an inner casing is lowered into place. The space between them is filled with cement grout.

Gravel pack

Gravel pack is used in formations composed of fine-grained soils having uniform grain size. As illustrated in Figure 2-16, a bed of gravel is installed around the screen, which in effect gives greater surface area for the infiltration of water into the well, while effectively blocking the entrance of sand.

The most common construction method is to install a large-diameter casing into the water-bearing strata and then lower a small casing with a well screen into the hole. The area around the screen is then filled with gravel as the outer casing is withdrawn corresponding to the length of the screen. The gravel used must be clean, washed, and composed of well-rounded particles that are four to five times larger than the median size of the surrounding natural material. The size and gradation of the gravel are critical in effectively blocking the entrance of fine sand.

Well screens

Wells completed in unconsolidated formations such as sand and gravel are usually equipped with screens (Figure 2-17). A properly sized screen allows the maximum amount of water from the aquifer to enter the well with a minimum of resistance, while blocking formation materials from passing into the well. Well screens are made from a variety of materials, including plastic, mild steel, red brass, bronze, and stainless steel. Types of well screens include sawn or slotted, wound, and bridged and louvered. The type of screen and material used depends on the type of soil, corrosivity of the water, cleaning and redevelopment methods, and other factors.

The size of a screen, or the slot number, is usually expressed in thousandths of an inch or in millimeters. The slot size is usually selected to permit some of the formation material to pass through it, depending on the uniformity of the grains. During development of the well, the fines pass through the screen while larger particles are held back, forming a graded natural-gravel barrier around the screen (Figure 2-18). The gravel pack is placed between the screen and the formation to help stabilize the formation materials.

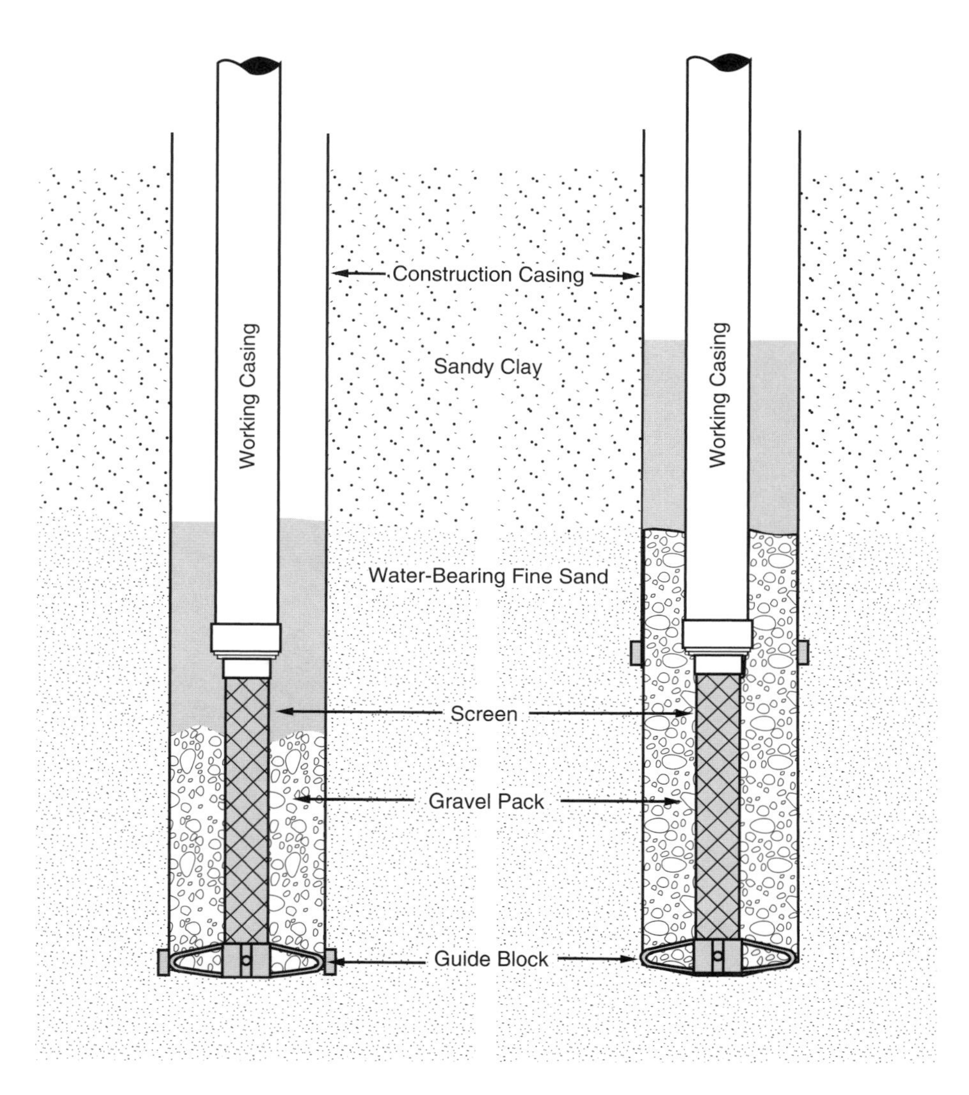

FIGURE 2-16 Gravel-wall well construction

The open area of the screen after the well has been developed must be carefully estimated because up to 50 percent of the screen slots may be plugged by formation particles. After the slot size has been determined, either the screen length or diameter must be adjusted to provide the total required screen opening.

FIGURE 2-17 Properly designed well screen

Information and specifications for screen selection are furnished in AWWA A100, *Standard for Water Wells* (latest edition).

Grouting

Wells are cemented, or grouted, in order to

- Seal the well from possible surface water pollution
- Seal out water from water-bearing strata of unsatisfactory quality
- Protect the casing against exterior corrosion
- Restrain unstable soil and rock formations

After a well is drilled and its casing installed, space is left along the length of the well between the casing and the formation material. This cavity must be sealed to prevent contaminants from getting down the well, either directly from the land surface or through formations with crevices connected to the surface (Figure 2-19). In formations that tend to cave in, such as sand, the cavity tends to seal itself eventually as the material shifts against the casing; in stable formations, such as solid rock, the cavity may remain indefinitely open if it is not filled.

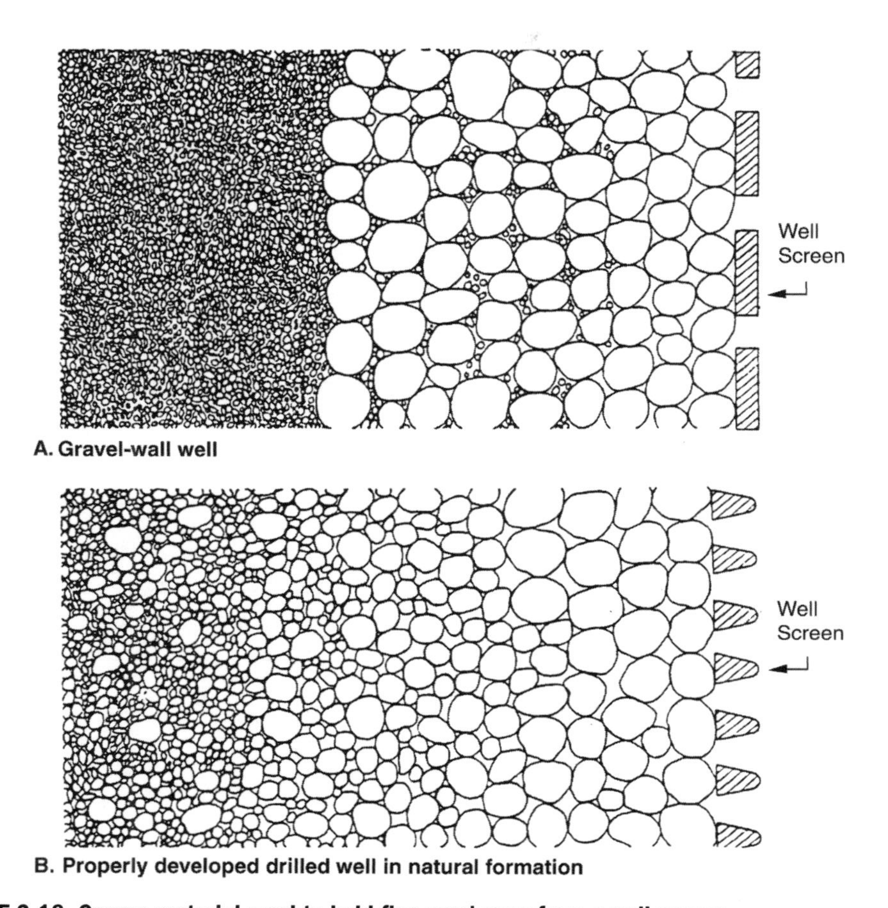

FIGURE 2-18 Coarse material used to hold fine sand away from a well screen

When corrosion of the casing is likely, a well is usually drilled larger than the casing so that the extra space can be filled with grout. Grout typically contains bentonite and portland cement. Portland-cement grout is usually mixed in a ratio of about one bag of cement to 5.5 gal (21 L) of clean water. Special additives are often used to accelerate or retard the time of setting and provide other special properties to the grout. Bentonite clay is used because it expands, thereby filling the voids around the casing.

The grout must be placed in the annular space, starting at the bottom and progressing upward so that no gaps or voids exist in the seal. Different methods may be used to place the grout depending on the size of the annular space, including a dump bailer, water-pressure driving, pumping, or a tremie pour. Grout must be placed in compliance with state regulations.

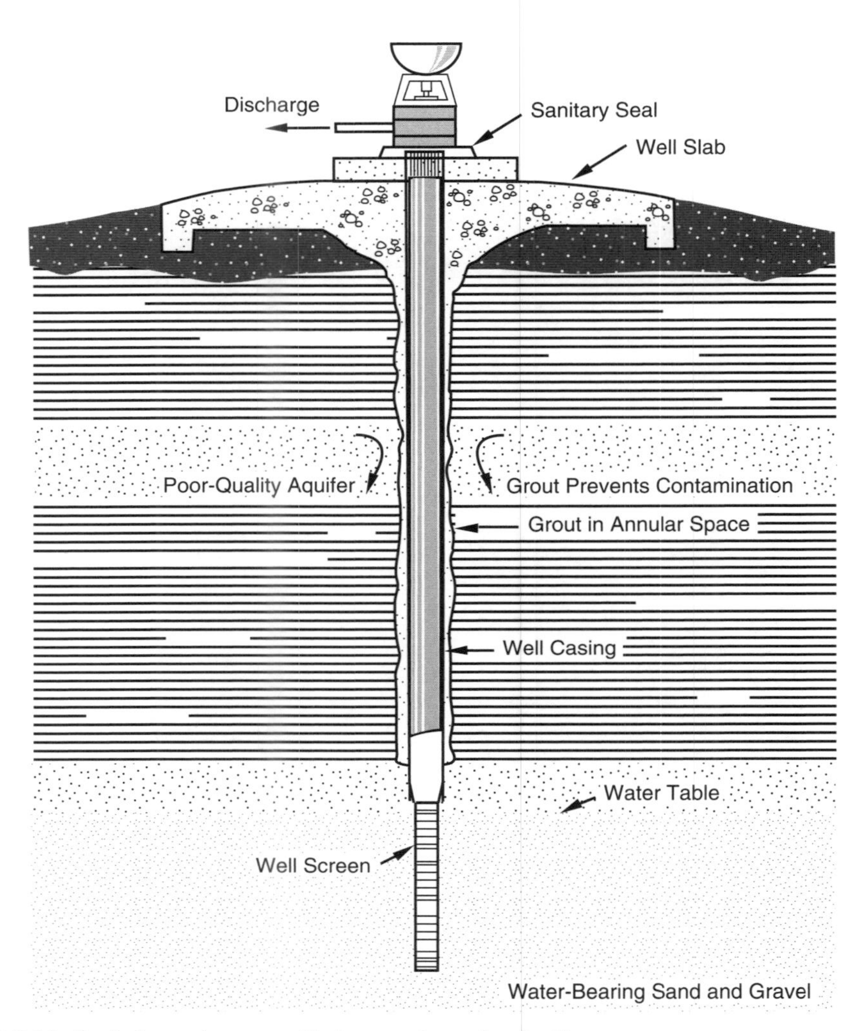

FIGURE 2-19 Sealed annular space that prevents contamination

Well Development

After a well is constructed, various techniques are used to develop it—that is, to allow the well to produce the best-quality water at the highest rate from the aquifer. The development method used is dependent on many factors and is usually selected by the well construction contractor. Methods commonly used include:

- High-rate pumping
- Surging
- Increased-rate pumping and surging
- Use of explosive charges
- High-velocity jetting
- Chemical agents
- Hydraulic fracturing ("fracking")

High-rate pumping

When a well draws water from a rock aquifer where no screen is used, development usually involves simply pumping the well at a rate higher than the normal production rate. This is done to flush out the casing and surrounding aquifer of fine material.

Surging

Surging generally employs a plunger, or surge block, on a tool string, which is moved up and down in the well casing (Figure 2-20). This creates flow reversals in the well face that dislodge the fine material and allow it to fall to the bottom of the well. The fines can then be removed with a bailer. Various types of surge blocks are used.

Surging is also achieved by use of compressed air. A seal is made at the top of the casing so that applied air pressure drives the water back through the screen or exposed walls of the well. A jet of compressed air is then used to raise fine material that has been freed up to the surface for removal. When this system is used, care must be taken not to drive air into the formation because it may collect in pockets that will reduce the rate of flow into the well.

Increased-rate pumping and surging

Increasing the pumping rate in combination with surging is effective in developing wells in unconsolidated limestone. As much of the drilling fluid as possible is first removed from the well and, as the water clears, gentle surging assists in development. Pumping rates are gradually increased in steps to develop the well fully.

Use of explosive charges

Explosive charges may be used in well development to fracture massive rock formations and create fractures that allow water to flow. Heavy charges are set off in the well bore at predetermined points to radiate fractures outward from the bore. Another technique is to shoot dynamite in the most permeable portions of a formation to clear drilling mud or other materials that are plugging the bore face. A third method is to set off very light charges in a series of explosions to produce vibrations that agitate the materials surrounding the bore. With any of these explosive methods, the loosened material must be removed from the well, usually by bailing.

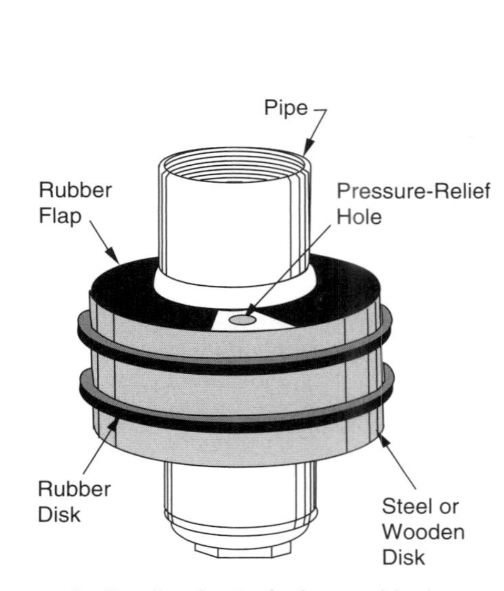

A. Details of a typical surge block

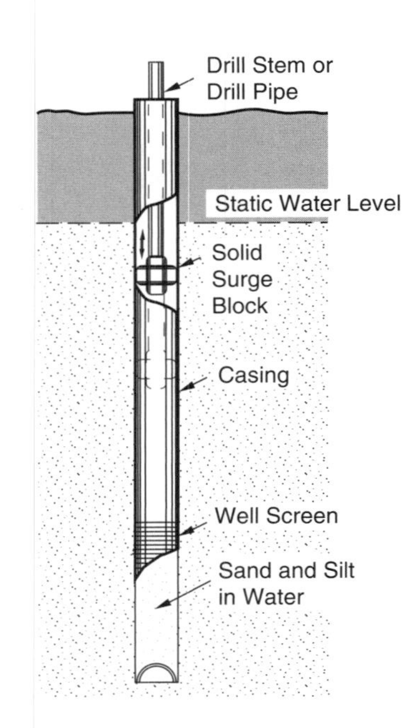

B. Surge block being used to develop a well by pulling silt and fine sand through the screen

FIGURE 2-20 Surge block schematics

High-velocity jetting

Hydraulic jetting involves high-pressure pumping of clean water against the sides of the well. When the force strikes the formation, it breaks down dense compacted materials and mud left from rotary drilling. The jets are rotated at intervals of from 6 to 12 in. (150 to 305 mm) throughout the water-entry area of the well. The well is simultaneously pumped out at a rate higher than the rate at which jet water is being introduced.

Chemical agents

It has been found that chemical additives frequently enhance the effectiveness of various development methods. For instance, wetting agents used with hexametaphosphates are particularly useful in increasing the effectiveness of jetting. Acids may also be used where there are iron or carbonate deposits that must be dispersed.

Hydraulic fracturing

Hydraulic fracturing (fracking) is a process in which a fluid is forced into the well under pressure great enough to open the separations between strata and along existing fractures. Gelling agents and selected sands are added to the fluid to hold the fractures open after pressure is removed. This process can be used only in rock aquifers, and the well casing must be firmly sealed and anchored in place.

Pumping Tests

After a well is developed and the quality of water produced is satisfactory, pumping tests are performed. These practical tests are intended to confirm that the well will produce at its design capacity. Most water supply well contracts specify the duration of the drawdown test that must be conducted to demonstrate the yield of the aquifer.

Many well acceptance tests are conducted with a temporary pump installed, usually powered by a gasoline or diesel engine. The test period for public water supply wells is generally at least 24 hours for a confined aquifer and 72 hours for an unconfined aquifer.

A means of measuring the flow rate of the pumped water must be provided. A commercial water meter, an orifice or weir, or periodic checking of the time required to fill a container of known volume are all methods of measuring flow rate. The pumping rate must be held constant during the test period and the depth to the water level in the well must periodically be measured. Drawdown is usually measured frequently (often at 1-minute intervals) during the first hour or so of the test and is then measured at gradually longer intervals (often up to 10-minute intervals) as the test period progresses.

Sanitary Considerations

Whenever possible, wells should be developed from formations that are deep enough to protect the water from surface contamination. Where a shallow groundwater source must be used, surface water contamination can be reduced if an impervious surface is created over the ground surface surrounding the well. One method is to place a 2-ft (0.6-m) layer of clay having a radius of about 50 ft (15 m) around the well. Filling the space between the casing and the hole with cement grout prevents contamination of wells by surface water flowing along the exterior of the casing. A concrete cap or platform should then be put in place around the well and built high enough to prevent floodwater from entering the well. The construction methods for wells are detailed by state regulatory agencies and, in some cases, by local jurisdictions as well. The well driller must follow the requirements, and the installation must be approved before the well is certified for use.

It is almost impossible to completely avoid contamination of the soil, tools, aquifer, casing, and screen during construction, so the water from the well, even after development, is likely to be contaminated. Extended pumping will usually rid the well of contamination, but disinfection with a chlorine solution is often quicker and more desirable.

A well is typically disinfected by the addition of sufficient chlorine to give a concentration of 50 mg/L. The pump can then be started and stopped to surge the disinfectant into and out of the aquifer, throughout the length of the well, and through the pump components. It is difficult to disinfect gravel packing by this procedure, so powdered or tablet calcium hypochlorite is usually added when the gravel is being placed.

For more details and specific procedures, refer to AWWA C654, *Standard for Disinfection of Wells* (latest edition).

AQUIFER PERFORMANCE

The sustained use of a well depends on the performance of the aquifer as a whole. Aquifer parameters also define how pumping in one well will affect other wells. These parameters are generally obtained via the pumping tests performed after well construction is complete. To measure changes in an aquifer during the pumping tests, small-diameter test wells called observation wells are installed. There is no set number of observation wells that should be used. One well positioned near an operating well is sometimes sufficient, but several may be required under different circumstances.

Aquifer Evaluation

The evaluation of an aquifer requires precise knowledge of the location of observation wells. The distances measured between the well being tested and the observation wells—and between the observation wells themselves—should be established carefully. The water level in each observation well is periodically measured relative to an elevation reference point, or benchmark.

One method of measuring the water depth in a test well is to use a steel tape graduated in tenths and hundredths of a foot (or metric units), with a weight attached (Figure 2-21A). The tape is chalked and lowered into the well until the weight reaches the bottom. When the tape is withdrawn, the water depth is indicated by the wetted position on the tape. Electronic depth-measuring devices are also available. They include a probe that is lowered into the well to activate a signal when water is contacted (Figure 2-21B).

Another way of measuring water depth, effectively used for many years for fixed installations, is the air-pressure tube method (Figure 2-21C). A small-diameter tube is suspended in the well beneath the water level, and the exact distance from the bottom of the tube to the ground surface is known. The top of the tube is fitted with a pressure gauge and a source of compressed air. Determining the water depth involves forcing air into the tube until it bubbles from the submerged end of the tube. The pressure gauge is then read, and the submerged length of tube can be calculated from the pressure required to displace the column of water in the tube. The calculation method is detailed in *Basic Science Concepts and Applications*, part of this series.

Evaluating Performance

Aquifer performance can be evaluated by three methods:

- Drawdown
- Recovery
- Specific-capacity

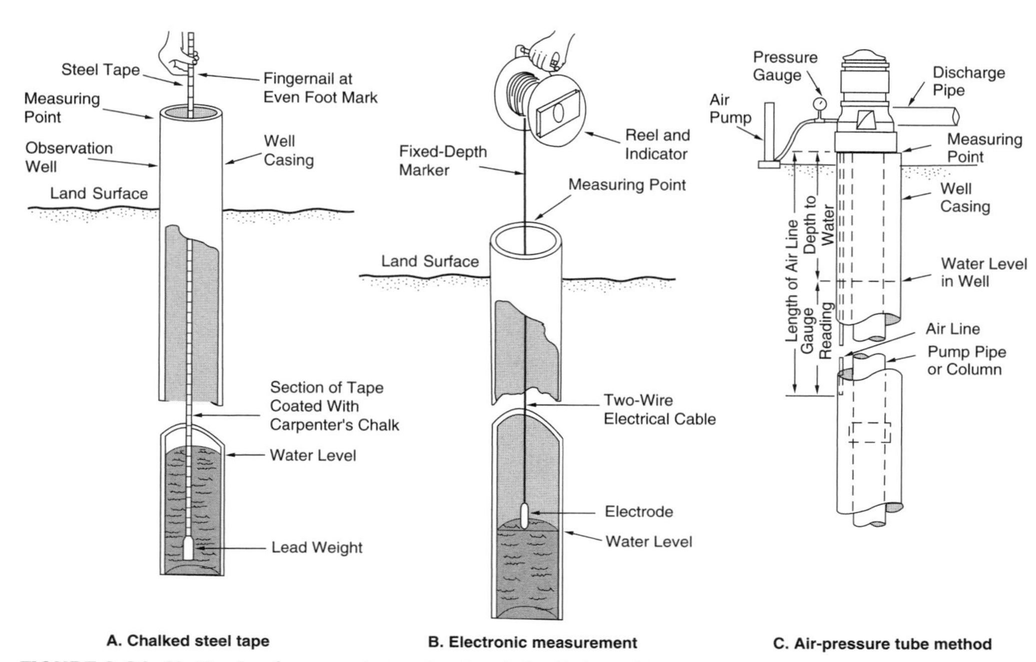

FIGURE 2-21 Methods of measuring water level depth in wells

In the drawdown method, the production well is pumped and water levels are periodically observed in two or more observation wells. The data are plotted and can be analyzed by various methods to relate drawdown in feet (or meters) to time measured in hours or days at a specific pump rate.

The recovery method involves measuring the change in water level in an observation well after the pumping has been stopped.

The specific-capacity method involves a relatively short test. As discussed earlier, specific capacity is the well yield per unit of drawdown. It does not indicate aquifer performance as completely as the other tests. However, it is especially valuable for evaluating well production after a period of time and for making comparisons with the data from when the well was new.

WELL OPERATION AND MAINTENANCE

When a well is in service, it is important to maintain and monitor records for any performance changes that might indicate future problems. Data from the following tests should be recorded regularly for each well:

- Static water level after the pump has been idle for a period of time
- Pumping water level

- Drawdown
- Well production
- Well yield
- Time required for recovery after pumping
- Specific capacity

Conditions for these tests should be the same each month so that direct comparisons can be made.

Regular maintenance of all the structures, equipment, and accessories is important to provide long-term, trouble-free service. Wells can fail if the casing or screen collapses or corrodes through. Perhaps the most common operational problem for wells is the plugging of the screen. The causes of screen plugging can be mechanical, chemical, or bacteriological. It is usually best to get professional assistance to correct the problem. If the screen is accidentally damaged, repair can be extremely expensive; the entire well may have to be replaced.

Periodically, bacteriological samples should be collected directly from each well and tested. If there are indications of contamination, periodic disinfection of a well may be necessary to prevent growth of nuisance bacteria that can lead to production problems.

WELL ABANDONMENT

When test holes and production wells are no longer useful, their equipment should be dismantled and their shafts sealed before they are abandoned. Proper preparation of a test hole or well for abandonment will accomplish the following:

- Eliminate a physical hazard
- Prevent groundwater contamination
- Conserve the aquifer
- Prevent mixing of desirable and undesirable water between aquifers

The basic objective governing the proper sealing of abandoned wells is the restoration of the geologic and hydrologic conditions that existed before the well was constructed. Each well that is abandoned should be considered unique. The methods used should be those that will give the best results at that location. Recovery of pumps, casings, screens, and other hardware is best performed by the firm that installed the well. The well construction company is also likely to be the best firm to complete the abandonment procedures properly.

SPRINGS AND INFILTRATION GALLERIES

Springs and infiltration galleries are both types of conduits between groundwater and surface water. While it is important to recognize their hydrologic importance, they are not typically used as the water source for a public water supply because it is difficult to predict their reliability and more effort is required to protect them from contamination.

Springs

Springs occur where groundwater exits the ground surface (i.e., where the groundwater surface and ground surface intersect) and are typically found at the base of a hill. Artesian springs occur where water from a confined aquifer is under enough pressure to flow to the surface through fractures in the rock or soil. Springs generally take the form of free-flowing water from fractures in an exposed bedrock surface or embankment or as seeps in soil embankments.

Springs are difficult to protect from contamination because they are generally closely connected to surface water infiltration and because the water intake is at the ground surface. This is in contrast to deep wells, which are not as readily influenced by surface water and which have pumps with intakes at depths below where animals or humans can easily access. When a water supply intake is placed at a spring, it should be enclosed in a tamper-proof, vermin-proof concrete box, and the site should have diversion channels that route surface drainage around the area. Furthermore, the protective structure should be designed to prevent clogging by sediment, debris, or ice. Drains should be installed to prevent stagnation during times of low flow, and access ways should be built to allow for periodic inspection and cleanout. Operational and structural controls should be in place to keep chemicals a safe distance away from the spring and outside the spring's recharge area. Regulatory requirements for treatment and periodic chemical analysis should be determined. They tend to be more stringent and more costly for springs than for well sources.

Flow rates from springs are often strongly influenced by the time of year and the amount of recent precipitation. Drought conditions or climate changes that reduce precipitation over an extended amount of time may significantly reduce the rate of flow from a spring. So before a spring is depended on for water supply, adequate observations should be made to ensure that the spring is as reliable as it needs to be.

Infiltration Galleries

The purpose of an infiltration gallery is usually to collect water from a surface water source at a point where the water can flow down through a filter material, such as several feet or meters of sand or gravel, to remove most of the particulate matter. It is generally more costly to install infiltration galleries than surface water intakes, and they are used when the cost for primary treatment and infiltration of the surface water justifies the additional effort.

As illustrated in Figure 2-22, a typical infiltration gallery installation involves the construction of a trench that is parallel to a streambed and about 10 ft (3 m) beyond the high-water mark. Perforated or open joint pipe is placed in the trench in a bed of gravel and then covered with a layer of coarse sand. The pipe and gravel can be wrapped with a geotextile filter fabric for additional filtration and to prevent silt clogging of the gravel pack.

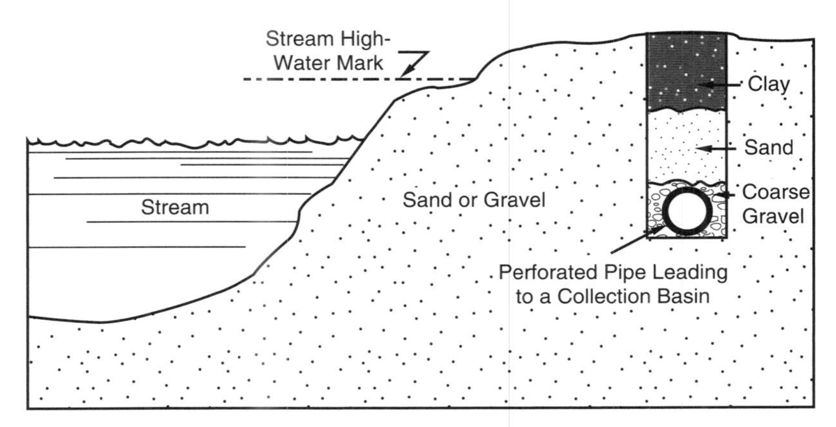

FIGURE 2-22 Example of an infiltration gallery

The upper part of the trench is filled with fairly impermeable material to reduce entry of surface water. The collection pipes terminate in a concrete basin from which water is pumped. Although the water collected in an infiltration gallery may be of much better quality than the surface source, it will not necessarily be free of all turbidity (cloudiness) or pathogenic (disease-causing) organisms. The water is therefore generally treated in compliance with state and federal requirements for surface water.

SELECTED SUPPLEMENTARY READINGS

AWWA Standard for Disinfection of Wells, ANSI/AWWA C654. Denver, Colo.: American Water Works Association (latest edition).

AWWA Standard for Water Wells, ANSI/AWWA A100. Denver, Colo.: American Water Works Association (latest edition).

Basics of Well Construction Operator's Guide. 1986. Denver, Colo.: American Water Works Association.

Bloetscher, F. 2009. The Impact of Unsustainable Ground Water, *Groundwater Protection Council Annual Forum (Salt Lake City, Utah)*. New York, NY: McGraw-Hill

Bloetscher, Frederick, and A. Muniz. 2008. Water Supply in South Florida—The New Limitations. *Sustainable Water Sources Conference Proceedings (Reno, Nev.)*. Denver, Colo.: American Water Works Association.

Bloetscher, Frederick, A. Muniz, and J. Largey. 2007. *Siting, Drilling, and Construction of Water Supply Wells*. Denver, Colo.: American Water Works Association.

Borch, M.A., S.A. Smith, and L.N. Noble. 1993. *Evaluation and Restoration of Water Supply Wells*. Denver, Colo.: Awwa Research Foundation and American Water Works Association.

Driscoll, F.G. 2007. *Groundwater and Wells*. St. Paul, Minn.: Johnson Screens, Inc.

Ekins, P. 2003. Identifying Critical Natural Capital—Conclusions about Critical Natural Capital. *Ecological Economics* 44:277–292.

Houben, G., and C. Treskatic. 2007. *Water Well Rehabilitation and Reconstruction.* New York: McGraw-Hill.

Hutson, S.S., N.L. Barber, J.F. Kenny, K.S. Linsey, D.S. Lumia, and M.A. Maupin. 2004. Estimated Use of Water in the United States in 2000. US Geological Survey Circular 1268, 46 pp. Reston, Va.: USGS.

Kresic, Neven, and Zoran Stevanovic. 2009. *Groundwater Hydrology of Springs*. Atlanta, Ga.: Elsevier.

Manual M21, Groundwater. 2003. Denver, Colo.: American Water Works Association.

Manual of Individual Water Supply Systems. 1982. US Environmental Protection Agency, Office of Drinking Water. EPA-570/9-82-004. Washington, D.C.: US Government Printing Office.

Manual of Instruction for Water Treatment Plant Operators. 1989. Albany: New York State Department of Health.

Nace, R.L. 1960. Water Management, Agriculture, and Ground-Water Supplies. US Geological Survey Circular 415, 12 pp. Reston, Va.: USGS.

Reilly, Thomas E., Kevin F. Dennehy, William M. Alley, and William L. Cunningham. 2009. Ground-Water Availability in the United States. USGS Circular 1323. Reston, Va.: USGS.

US Geological Survey. 2009. Estimated Use of Water in the United States in 2005. USGS Circular 1344. Denver, Colo.: US Geological Survey. Available: http://pubs.usgs.gov/circ/1344/.

US Geological Survey Ground-Water Resources Program. 2001. USGS Fact Sheet 056-01. Available: http://water.usgs.gov/ogw/.

Water Quality and Treatment, 6th ed. 2010. New York: McGraw-Hill and American Water Works Association (available from AWWA).

CHAPTER 3

Surface Water Sources

Surface water is the term used to describe water on the land surface. The water may be flowing, as in streams and rivers, or quiescent, as in lakes, reservoirs, and ponds. Surface water is produced by runoff of precipitation and natural groundwater seepage. In this book, surface water is defined as all water open to the atmosphere and subject to surface runoff. That definition distinguishes surface water from both groundwater and ocean water.

Most large population centers in the United States are supplied by surface water sources. Although groundwater is used as a water supply source in rural areas and by most small communities, it is rare for sufficient groundwater to be available to serve large cities. Many large cities, such as those around the Great Lakes and along many major rivers, have thrived because of the availability of fresh water. In the arid Southwest, water may be routed great distances to a city from other areas through canals or pipelines.

SURFACE RUNOFF

Patterns of surface runoff are described in terms of land areas that drain rainfall and melting snow to a certain point in a river or stream. Those land areas may be called drainage areas, drainage basins, catchments, or watersheds. Each watershed is bounded by relatively high ground called a divide, which separates it from other watersheds.

Groundwater from springs and seeps also contributes flow to most streams. This is often referred to as base flow. If the adjacent water table is at a level higher than the water surface of the stream, water from the water table will flow to the stream (Figure 3-1). If the water table is at a level lower than the water surface of the stream, the stream will recharge the groundwater. The pattern of flowing to or from the stream can vary according to seasonal fluctuation of the water table. Many streams would dry up shortly after a rain if it were not for groundwater flow.

Influences on Runoff

The following principal factors affect how rapidly surface water runs off the land:

- Rainfall intensity
- Rainfall duration
- Soil composition
- Soil moisture
- Ground slope
- Vegetation cover
- Human influences

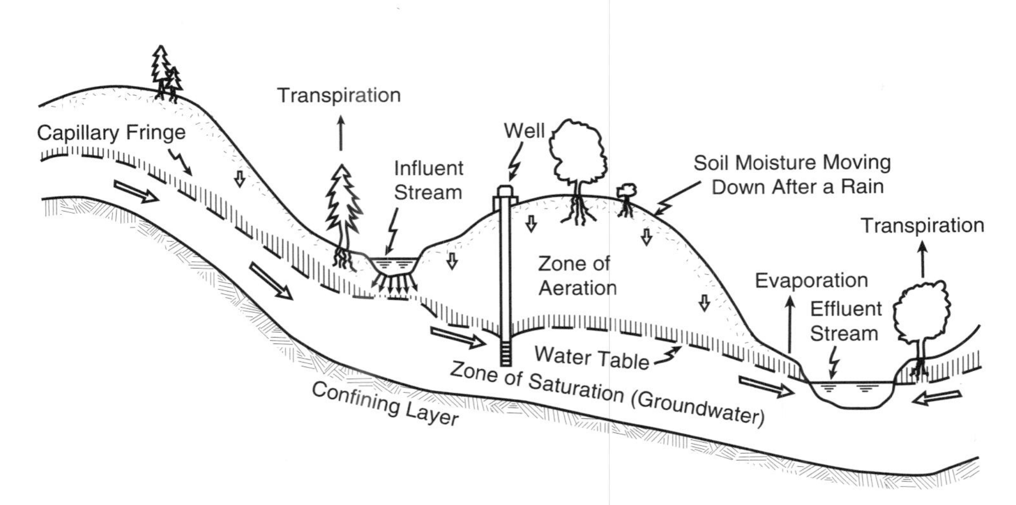

FIGURE 3-1 Groundwater and stream interaction
Adapted from Linsley et al. 1999.

Rainfall intensity

In a typical storm event, the intensity of rainfall is low at first, and the initial rainfall soaks into, or infiltrates, the ground (provided the ground has not been paved over or built upon). As rainfall intensity increases and the ground becomes wetter, however, the rate at which the rain falls will eventually exceed the rate at which the ground can absorb it, and runoff will be generated.

Rainfall duration

The amount of runoff increases with rainfall duration for two reasons: first, because as the soil gets wetter its capacity to absorb more rain decreases, and second, because after runoff begins, a portion of each additional increment of rainfall is converted to runoff.

Soil composition

The composition of the surface soil also significantly influences how much runoff is produced. Coarse sand, for example, has large void spaces that allow water to pass through readily, so even a high-intensity rainfall on sand may result in little runoff. On the other hand, even light rainfall on relatively impervious soils, such as those that contain a lot of clay, tends to result in a relatively large amount of runoff. Clay soils not only have small void spaces, but the soil swells when wet. This expansion closes the void spaces, which further reduces the infiltration rate and results in greater surface runoff.

Soil moisture

If soil is already wet from a previous rain, surface runoff occurs sooner than if the soil were dry. Consequently, the amount of existing soil moisture has an effect on surface runoff. If the ground is frozen it is essentially impervious, and runoff from rain or melting snow may approach 100 percent.

Note that although the increase in soil saturation means a sharp decrease in its capacity to absorb water (i.e., its infiltration rate), the saturation of a soil does not mean it has altogether lost its capacity to absorb water. Depending on the soil composition, a saturated soil's infiltration rate can still be significant.

Ground slope

Where land is relatively flat, runoff tends to spread out and soak into the ground more easily. Where land is sloped more steeply, however, runoff more readily collects in rivulets and flows downstream, resulting in a relatively high rate of runoff.

Vegetation cover

Vegetation plays an important role in limiting runoff. One way it does this is by the process of interception that has already been described in chapter 1.

Vegetation also influences runoff by the way it affects surface conditions and soil composition. At the surface, decaying organic litter such as leaves, branches, and pine needles readily absorbs water and helps move it into the ground. Vegetation and organic litter also act as a cover to protect soil from getting compacted or crusted over. The physical interception of rainfall by vegetation prevents raindrops from dislodging fine soil that can clog open channels at the soil surface or be transported downstream as sediment. Vegetative cover also reduces evaporation of soil moisture.

Decaying vegetative matter on top of the soil eventually becomes part of the soil. Soil that is high in such organic matter can absorb relatively large amounts of water while maintaining many openings through which water can pass to the soil layers below.

Human influences

Human activities can have a dramatic effect on rates and amounts of surface runoff. The impervious surfaces of buildings and of streets and other paved areas greatly increase the amount of runoff and the peak flow rates that occur during a storm. Drainage structures such as gutters, ditches, and canals have conventionally been designed to convey water efficiently downstream, diminishing the opportunity for water to soak into the ground and percolate to groundwater, where it can slowly supply the base flow of nearby streams. In general, increasing amounts of imperviousness in a drainage area result in flashy streams that are more likely to flood during storm events but become drier in between storms.

Stormwater detention facilities can help keep rates of runoff from becoming excessive immediately downstream, but they cannot reduce the total volume of runoff coming from upstream. Some states and local jurisdictions have passed stormwater management laws requiring that infiltration devices be used to lessen the effects of increased runoff from urbanization. A few jurisdictions are encouraging the use of site development techniques that reduce the amount of runoff generated. These development techniques are often referred to as sustainable urban drainage systems (SUDS), water-sensitive urban design (WSUD), low-impact development (LID), and environmental site design (ESD).

Watercourses

Surface runoff naturally flows along the path of least resistance. All of the water within a watershed flows toward one primary natural watercourse unless it is diverted by a constructed conveyance such as a canal or pipeline or it flows back into the ground to become groundwater.

Natural watercourses

Typical natural watercourses include brooks, creeks, streams, washes, arroyos, and rivers. Watercourses may flow continuously, occasionally, or intermittently, depending on the frequency of rainfall, the availability of snowmelt, and the rate at which groundwater contributes to the base flow of the watercourse.

Perennial streams. Natural watercourses that flow continuously at all times of the year are called perennial streams. They are generally supplied by a combination of surface runoff, springs, groundwater seepage, and possibly snowmelt.

Ephemeral streams. Streams that flow only occasionally are called ephemeral streams (Figure 3-2). These streams usually flow only during and shortly after a rain and are supplied only by surface runoff.

Intermittent streams. The frequency of flow in intermittent streams falls somewhere between perennial and ephemeral. The streams flow when groundwater levels are high, and they are dry at other times. Depending on the amounts of rainfall and snowmelt and the resulting groundwater levels, intermittent streams may flow for weeks or months at a time and then remain dry for awhile.

Constructed conveyances

Many facilities are constructed to hasten the flow of surface water or to divert it in a direction other than the one in which it would flow under natural conditions. Ditches, channels, canals, aqueducts, conduits, tunnels, and pipelines are all examples of constructed conveyances. In many cases, constructed conveyances are used to divert water from its natural watershed into another watershed where it is needed (referred to as inter-basin transfer).

Many local conveyances are constructed to prevent water from ponding in areas where it would impede agricultural or urban development. Other examples of constructed

FIGURE 3-2 Bed of an ephemeral stream

conveyances include canals provided for boat access or shipping, aqueducts to provide irrigation water to arid regions, and pipelines to bring potable water to areas that have an insufficient natural supply.

Impoundments

For the purposes of this discussion, an impoundment is defined as any significant accumulation of water in a basin surrounded by land. The basin can be a natural formation, such as a pond or lake, or a reservoir constructed to serve some specific water need. Reservoirs range from small basins constructed by farmers for watering livestock to massive reservoirs that store water for public water systems and recreational uses.

SURFACE SOURCE CONSIDERATIONS

The development of a surface water source for use by a public water supply requires careful study of the quantity and quality of water available and of the expected rates of water demand.

Quantity of Water Available

The prime consideration in selecting a water source is that the supply must reliably furnish the quantity of water required. The flow in any watercourse fluctuates in relation to the amount of rainfall and runoff that occurs, so it is crucial to know that a sufficient amount of water will be available for withdrawal during low-flow or drought conditions.

Community water needs are normally the greatest during the warm months of summer and early fall, and this is usually the period when natural streamflow is the lowest. To compensate for this problem and ensure water supply reliability, reservoirs can be constructed to store water during periods of excess supply. The stored water can then be tapped during periods when natural streamflow is deficient.

Safe yield is a concept that describes the availability of water through a critical dry period. It is calculated based on the size of the impoundment, the area of the watershed, and the desired level of reliability. In simple terms, the safe yield represents the maximum rate at which water can be withdrawn continuously over a long period of time, including during very dry periods. For example, without storage, the safe yield of a stream is simply the amount of water that can be withdrawn during a period of lowest flow. In many states, the safe yield of a direct stream withdrawal for public water supply is the "1Q30," or the lowest one-day flow that occurs on average once every 30 years.

Safe yield from a natural water source can be increased by adding storage to the water supply system. If a stream has a high enough flow in the winter or other season to replenish water in an impoundment, the amount of water available for withdrawal during dry seasons can be greatly increased.

Many interrelated factors must be considered when calculating the safe yield of a public water supply. In addition to rainfall and streamflow considerations, allowances must also be made for other uses for agricultural irrigation or by other water systems. The amount of instream flow required to sustain downstream habitat for fish and other aquatic organisms is also an issue that must not be overlooked.

Water Quality

Although it is technically possible to treat water of just about any quality to make it suitable for a public water supply, it is often not economically practical to do so. In some situations, it may be more cost effective to pipe water a considerable distance from a remote, good-quality supply than to treat poor-quality water that is available locally. In some parts of the United States, for example, it is less expensive for some coastal cities to transport fresh water from sources hundreds of miles away than to desalt the ocean water nearby. Elsewhere in the world, however, desalinization must be used because the amount of available fresh water is not sufficient.

The economic concerns associated with drawing on a limited water resource to provide adequate water where several communities need it are often complex. Costs are influenced by the supply and demand among all the parties that have an interest in using the water. Political and legal considerations, such as who has the rights to the water, also come into play. The costs of the energy needed to operate the treatment plant may fluctuate considerably over the life of the plant.

As water treatment technology becomes more efficient, however, the cost of treating poor-quality sources decreases, making treatment a more attractive option for some communities.

Some of the principal quality factors that must be considered in evaluating the suitability of a water source for use are

- excessive turbidity,
- microbiological contamination,
- chemical or radiological contamination,
- undesirable taste, odor, or color,
- presence of algae growth, and
- water temperature.

The influences of factors on water use and treatability are covered in *Water Treatment* and *Water Quality*, other titles in this series.

WATER STORAGE

Water can be stored by natural features, artificial impoundments, or deliberately recharging natural aquifers.

Natural Storage

Natural lakes are often used as a source for public water supply. Natural lakes occur most often in glaciated regions. Large lakes such as the Great Lakes of North America were formed by the movement of massive lobes of ice that left deep, wide valleys after the glaciers receded. The melting of large ice blocks that had been earlier stranded by the receding glaciers formed the smaller lakes (called *kettles*) in that region.

Generally, a lake must be relatively isolated from human activities to have good water quality. Lakes having a watershed with heavy agricultural activity, industrial operations, or urban development are often highly polluted.

If a lake located some distance from a community is being considered for use, it must be decided whether it is better to treat the water at the source and then pipe the finished water to the community or to pipe the raw water to the community and treat it there. Because of the considerable cost of piping water over long distances, an engineering analysis of all the alternatives must be made.

Impoundments

An impoundment is a lake that has been created either by carving out a basin or by building a dam across a stream valley. Although there is evidence of dam construction as far back as 4000 BCE (Nile River), dams were not widely used until the mid-1800s. Dams were first used primarily for water supply and to harness river flow for mills. In glaciated regions, dams and perimeter berms have been used to increase the depth and storage capacity of kettle lakes.

With the development of modern construction equipment and techniques, dams have become a common way of creating a large reservoir. They are usually built where damming a valley stream or river can form the reservoir. The deeper and narrower the valley, the more economical it is to construct the dam. Dams can be made of earth, rock, or concrete.

Offstream reservoirs can be constructed next to major rivers to add storage to a water system. Water can be pumped from the river to the reservoir during high-flow periods, then withdrawn for water supply during low-flow periods. This type of impoundment is commonly referred to as a pumped-storage reservoir and can involve constructing a dam across the stream of a small tributary, thereby avoiding many of the adverse environmental and navigational effects of building a dam on a major river.

Types of dams

Dams are classified by what they are made of and how they resist the pressure forces imposed on them. The four general types of dams are gravity, arch, buttress, and embankment (Figure 3-3). The first three types are typically constructed of concrete; embankment dams are constructed either of earth-fill or rock-fill. A gravity dam resists both sliding and overturning by its own weight. In a buttress dam, which is a type of gravity dam, buttresses support reinforced concrete slabs, and the impounded water's weight is used to help resist sliding and overturning. In contrast, an arch dam resists movement by transferring the thrust of water pressure laterally to the rock walls of a canyon.

Dams constructed of earth and rock, called embankment dams, are the most common. Although the concept of an embankment dam seems simple, such a dam must be carefully engineered to resist the forces of the reservoir water and prevent damage or failure due to seepage or undermining.

A cross section of a typical earth dam and intake tower is shown in Figure 3-4. An embankment dam may be constructed mostly of sandy or rocky soils if they are locally available, but there must be an inner core of impermeable material to ensure watertightness. Earth dams often require an internal drainage system (using sand and gravel) to reduce water pressure in the dam, which in turn helps prevent soils from shifting or becoming soft because of the presence of water.

Rock-fill dams are another type of embankment dam. Rock-fill dams use a broad range of rock sizes to provide stability; watertightness is provided by building a reinforced concrete slab on the upstream face of the dam.

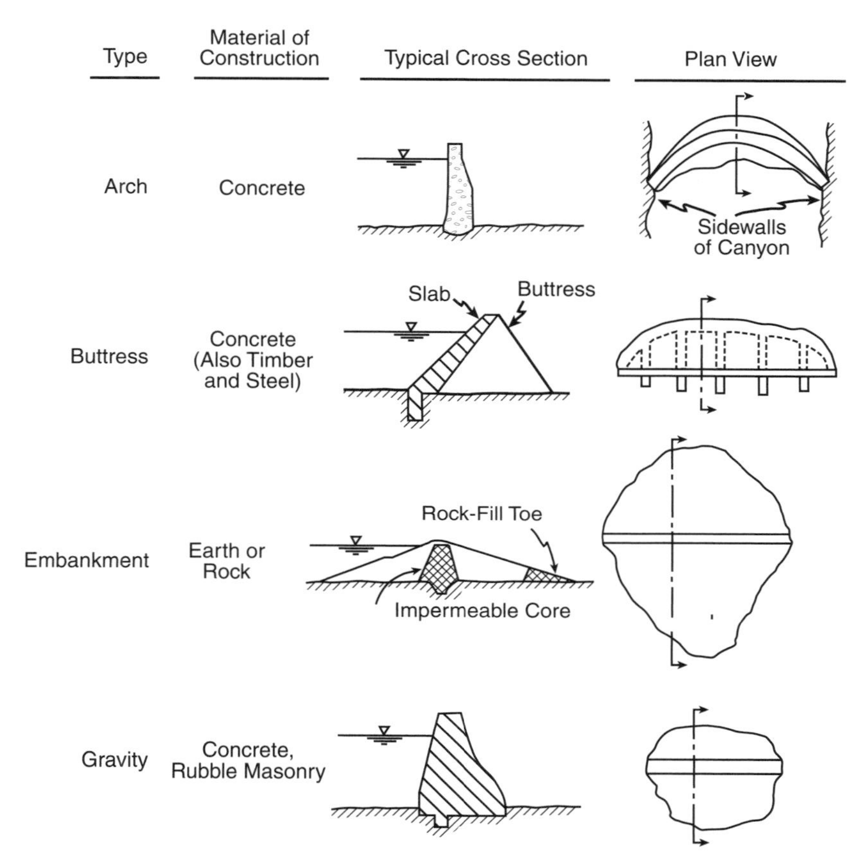

FIGURE 3-3 Types of dams
Adapted from Linsley et al. 1999

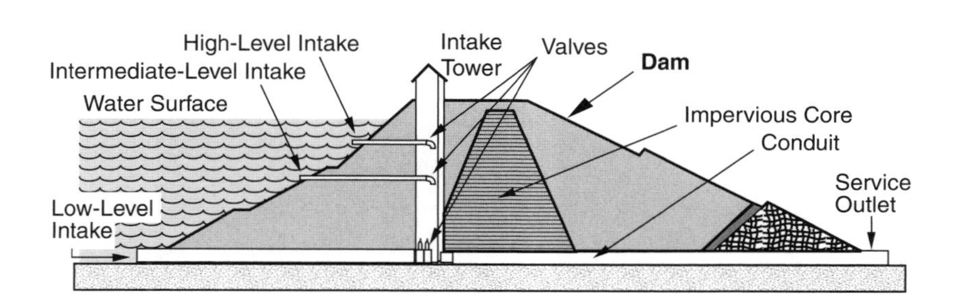

FIGURE 3-4 Cross section of dam and intake tower for impounded surface water supply

Dam design

Subsurface conditions are among the most important to consider in dam construction. The natural substrate on which the dam is constructed must have sufficient strength and be impermeable enough that the dam will not fail because of undermining by water. Subsurface conditions are investigated using core borings. If necessary, unsafe material must be removed and replaced with suitable material before the dam is constructed.

To keep construction costs to a minimum, a dam should be built with locally available materials to the extent possible. An earth dam is typically used where there is an abundance of fill material and where bedrock is very deep. A concrete dam is often used where fill is scarce and/or bedrock is shallow.

Some small reservoirs have sufficient capacity to hold all the water that flows into them, so no outlet is necessary. However, because most dams are constructed across a stream or river, they are expected to overflow at least occasionally, especially during flood conditions. Spillways must be provided to ensure that excess water will not rise to the point where it overtops and erodes the dam, but is conveyed downstream safely. If a dam is overtopped, the entire structure can be severely damaged or destroyed, depending on the construction of the dam and the degree of overtopping.

For most spillways, water simply flows freely over the spillway crest when the pool rises above that level. The spillway crest typically defines what is called the normal pool condition of the reservoir. Figure 3-5 shows a concrete chute spillway on the abutment of an earth dam. Other spillways consist of gates that must be raised to allow water to flow through the dam (Figure 3-6).

To ensure public safety, dams for water supply should be designed to withstand at least a 100-year flood event, defined as the rate of flow that has a 1 percent chance of being equaled or exceeded at that location in any given year. However, if failure of the dam could cause loss of life to people downstream, it should be designed for a larger flood. In many states, such dams must be designed to pass the probable maximum flood, which can be up to five times the size of a 100-year flood. Nearly all states have safety regulations that pertain to dam design and construction.

FIGURE 3-5 Yellow Creek Dam. Intake structure is to the left of the dam, and chute spillway is to the right of the dam.

FIGURE 3-6 Chute spillway on a concrete dam

Dam maintenance

Most of the larger dams in the United States have been constructed with federal funds and remain under federal ownership. Others are owned by the state or local jurisdiction or solely by a water utility.

When a dam is owned by a water utility, the system operator may have responsibility for dam maintenance. The structure must be inspected periodically and maintained to prevent failure. Failure of a dam can result in loss of a water source, destruction of downstream property, and injury or loss of life among the people living and working in that area. Water-control structures such as mechanical spillways and slide gates require periodic maintenance to keep them in good working order and to prevent failure.

Detailed records should be kept carefully of all inspections and work done on dams and water-control structures. Most states have specific inspection procedures and reporting requirements. Any inspection and maintenance procedures should be documented in an operation and maintenance manual.

The National Inventory of Dams (NID), maintained by the US Army Corps of Engineers, now lists approximately 75,000 dams in the United States. Because of the numbers of dams already built and the economic and environmental considerations of new projects, the rate of construction of new dams has declined. Some existing dams have even been destroyed because it has become more desirable to have a free-flowing river where the dam was built.

Groundwater recharge

A few attempts have been made to provide water storage underground by recharging depleted groundwater aquifers with surface water. One way to accomplish this is to excavate large basins that collect water and allow it to infiltrate into the subsurface. Direct injection of water into the subsurface is another approach.

The use of underground storage presents some technical challenges. If untreated surface water is to be injected into the soil, care must be taken not to pollute the aquifer. In addition, turbidity in the injected water, which is caused by fine suspended particles, will gradually plug the injection zone in the target formation(s). So to ensure that the storage system will continue to function as needed, the injection water must generally be treated to drinking water quality before it is stored. Because aquifers are not sealed off from the rest of the natural environment, losses can occur, and the volume of water pumped into the aquifer cannot always be fully recovered.

The use of underground storage brings important political and economic issues as well. It is not uncommon for the construction of new surface reservoirs to face opposition for a variety of environmental reasons. In some cases, projects may have been approved by local regulatory agencies but are then disapproved by regional, state, or federal agencies. When storage of large quantities of water is necessary but the creation of new surface impoundments is unacceptable, the options will be either to pipe water for long distances from where a reservoir is available or to store water in local aquifers. All injection of fluids into the ground (including water that has been treated to drinking water quality) is subject to federal and state approval under the rules of the federal Underground Injection Control (UIC) Program.

INTAKE STRUCTURES

Water is drawn into a water supply system from a surface impoundment or stream through an intake structure. Types of intake structures include simple surface inlets, submerged intakes, movable intakes, pump intakes, and infiltration galleries.

Surface Intakes

A simple surface intake consists of a small concrete structure situated near the bank of a river or lake. When a slide gate or stop log is removed from the opening, water spills through the intake into a canal, ditch, or pipeline, which then carries the water to a treatment plant. Larger intake structures may be built with a bar screen and have mechanical devices for removing accumulated trash from the screen.

The principal disadvantage of using a surface intake on a river or reservoir is that the water quality is not usually as good there as it is at greater depths. The water may be warmer in summer; have ice on the surface in winter; and may have algae, leaves, logs, and other floating debris at the surface at certain times of the year. The elevation of the water in the lake or stream may fluctuate substantially; therefore, the intake must be designed and located so that it will draw water under all conditions.

Submerged Intakes

Submerged intakes draw water from below the surface. The best water quality in most lakes and streams can usually be obtained in deep water. Intakes should not be located directly on the bottom or they will draw in silt and sand; therefore, they are often raised a short distance above the bottom. Submerged intakes have the advantages of presenting no surface obstruction to navigation and being less likely to be damaged by floating debris and ice. Many designs are used for submerged intakes; one typical type is illustrated in Figure 3-7. In some instances, screens with very fine openings are used to prevent fish from being drawn into the intake (Figure 3-8).

Many surface water treatment plants that normally draw water from a submerged intake also have an emergency surface intake (also called a shore intake). In the event of

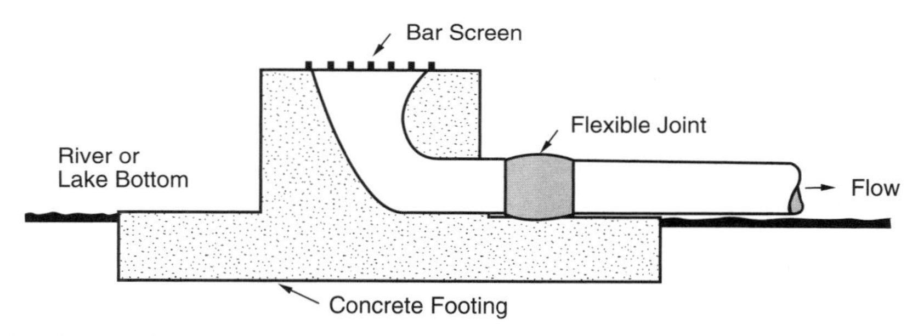

FIGURE 3-7 Schematic of a typical submerged intake

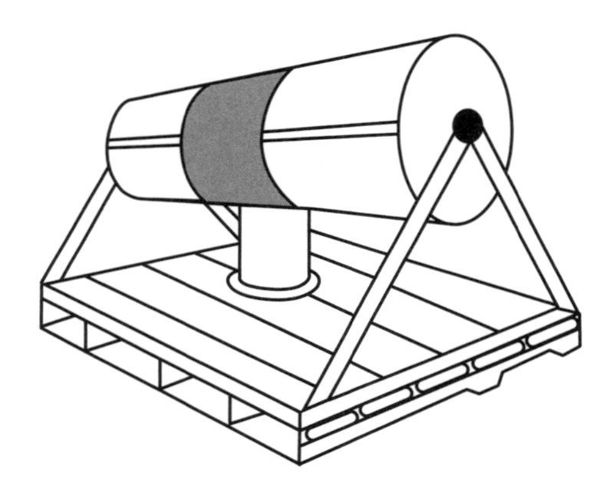

FIGURE 3-8 Fish screen for a submerged intake

some kind of clogging or other failure of the submerged intake, the surface intake can be activated to furnish water of usable quality while repairs are made.

Movable Intakes

Water quality at different depths can vary with the season depending on the degree to which the source lake or reservoir is subject to stratification and turnover during the year. During the summer months, warm water that is near the surface tends to stay near the surface because it remains less dense than the water below. If it contains sufficient nutrients, algal growths that taint the water may occur. In that case, water drawn from deeper levels is generally more likely to be of higher quality.

Near the bottom of the lake or reservoir, however, a different kind of water quality problem may occur. At that depth, low oxygen levels may result in conditions that tend to dissolve nutrients and metals that would otherwise remain in the bottom sediment. In that case, it is more desirable to draw water from above that lowest zone.

Finding the depth at which the best water quality is found is complicated by the phenomenon of turnover, which may occur once or twice a year as water temperatures change and the speed of wind at the surface changes. During turnover, the layers of water mix from top to bottom.

To ensure that the raw water that supplies the public system is withdrawn from a depth where the highest water quality is expected, movable intakes are frequently used. The use of movable intakes is generally advantageous in lakes and streams where the water levels vary greatly or there are other reasons for varying the depth of withdrawing water. Movable intakes are also used where good foundation is lacking or where other conditions prevent the construction of a more substantial structure.

Multiple-port Intakes

Intake structures that have multiple inlet ports are also used in locations where the water level or water quality varies (Figure 3-9). By opening and closing valves, the operator can withdraw or mix water from several different depths to compensate for changing water elevation, to avoid ice cover, or to select the depth from which the best-quality water can be withdrawn.

Pump Intakes

Pump intakes are used where it is not possible to construct a facility that will allow water to flow from the source by gravity alone. One design, illustrated in Figure 3-10, has an intake pump and suction piping mounted on a dolly that runs on tracks. As the water level rises or falls, sections of discharge piping are removed or added and the railcar is moved accordingly.

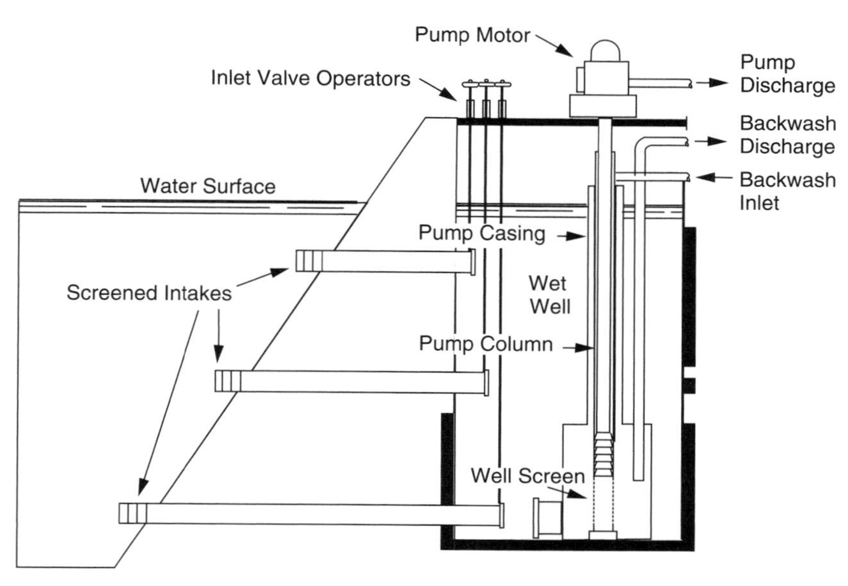

FIGURE 3-9 Intake structure with multiple inlet ports

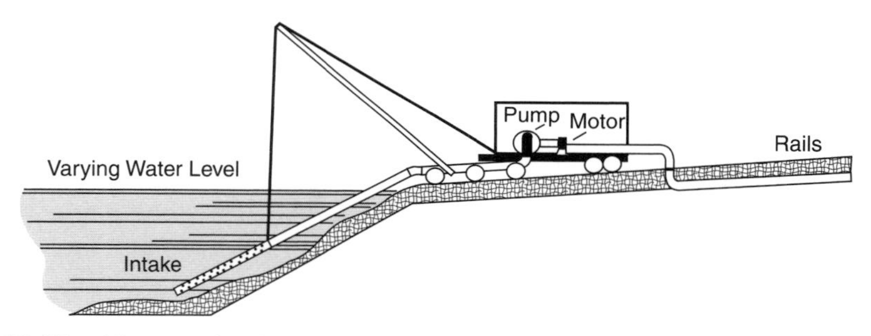

FIGURE 3-10 Movable pump intake

Infiltration Galleries

An infiltration gallery is a type of intake that is usually located next to a lake or river where it can pick up seepage from the surrounding sediment. A buried intake (instead of an open intake) can similarly be located in the bottom of a lake or stream, but it can be used only if the bottom sediment is porous enough and has the proper gradation to provide a reasonable flow rate. There must also be enough wave action or flow across the bottom to prevent the accumulation of fines, which would plug the sediment surface.

Where the proper conditions are present for installing an infiltration system, infiltration galleries usually provide water of much better quality than water taken directly from the source because the natural straining action of the sediment eliminates debris and reduces turbidity. Infiltration galleries also have the advantages of not freezing or icing over and they are not as vulnerable to damage.

SURFACE-SUPPLY OPERATING PROBLEMS

Surface water sources present several physical, chemical, and biological challenges that can be troublesome for the water system operator.

Stream Contamination

The quality of water taken from streams and rivers is influenced by all upstream water use and land use. The quality of water in major rivers that are used for navigation and by industries can decrease suddenly because of spills from barges, leaks from tank facilities, broken pipelines, accidental industrial spills, and overflows from combined and sanitary sewers. Utilities that draw water from particularly vulnerable sources depend on warning systems to give advance notice when pollutants are approaching their intake. Some streams also have occasional periods when water quality is especially poor as a result of natural causes, such as heavy spring runoff. Drought conditions can greatly diminish water quality because pollutants in the water (e.g., from wastewater effluent) are less diluted.

Any water system that is vulnerable to episodes of poor-quality water should construct adequate storage facilities. These facilities are then used when the intake is shut down due to poor-quality water at the intake point.

Lake Contamination

Lakes and reservoirs are also vulnerable to natural and human contamination. One problem encountered by some water systems using reservoirs in agricultural areas is the occasional occurrence of nitrate levels exceeding state and federal standards. High nitrate concentrations are caused by runoff from farmlands in the drainage basin, and they may persist for several months each year. High sediment loads carrying phosphorus, a common nutrient, are also a problem in agricultural basins and those undergoing urbanization.

Excessive growth of algae and aquatic weeds in reservoirs is quite common, particularly in warmer climates. High levels of nutrients in the water usually cause this, and the problem can be minimized by proper treatment and land use management.

Invasive aquatic species can present another threat. In the Great Lakes the zebra mussel, which was brought by international shipping vessels, is notorious for the problems it can cause. This organism grows profusely in intake structures and can quickly choke off the flow of water if it is not controlled. Another mussel, the quagga, causes similar problems. Unfortunately, these organisms are gradually being transported to other surface water impoundments by boats and birds. They will probably become established as permanent nuisance organisms in most waters of North America.

Control of aquatic plants, prevention of lake stratification, control of invasive mussels, and other treatment of source water are detailed in *Water Treatment*, also part of this series.

Icing

Operators in cold-weather areas often face additional operational challenges presented by frazil ice or anchor ice. When surface water is almost at the freezing point and is rapidly being cooled, small, disk-shaped frazil ice crystals form and are distributed throughout the water mass. When the frazil crystals are carried to the depth of a water intake, they may adhere to the intake screen and quickly build up to a solid plug across the opening.

Anchor ice is different in that it is composed of sheetlike crystals that adhere to and grow on submerged objects. Although experience varies at different locations, most systems experience ice blocking only at night. These conditions are generally favorable to ice crystal formation:

- Clear skies at night (because there will be high heat loss by radiation)
- Air temperature of less than 19.4°F (–7°C)
- Daytime water temperature not greater than 32.4°F (0.2°C)
- Winds greater than 10 mph (16 kph) at the water surface

A falling water level in the intake well usually indicates icing of an intake. When icing is noticed, it is generally best to stop using the intake immediately, because the blockage will only get worse. If sufficient storage is available or another intake can be used, the easiest method of ridding the intake of ice is just to wait; the ice will float off within a few hours.

Water utilities that frequently experience icing have tried providing the intake with piping to backflush the line with settled water or blowing the line with compressed air or steam. If more than one intake is available, alternating intakes allow any accumulated ice to melt and float away from the intake opening.

Ice formation at inlet structures can be minimized by using very widely spaced screening, or no screen at all, and by constructing the intake pipe and structures of nonferrous materials such as fiberglass. Some intake designs keep the entrance velocity very low by using large bell-shaped structures or several inlets.

Stream source intakes are not ordinarily deep enough to allow water to be taken from a depth beneath the supercooled level near the surface. Consequently, multiple intake points and use of heating elements may be necessary to prevent icing.

Evaporation and Seepage

Surface water reservoirs naturally lose water to evaporation and seepage. Rates of evaporation loss can range from less than 2 ft (0.6 m) per year in the northern part of the United

States to more than 6 ft (2 m) in the southwestern United States. To minimize evaporation losses from the surface of a reservoir, deep and narrow canyons are preferred over shallower impoundments. Seepage losses occur through the bottom and sides of the impoundment and can vary greatly depending on local soil type, geology, and the elevation of the lake surface relative to the surrounding groundwater table. These losses cannot economically be controlled in large reservoirs, so the reservoirs are generally designed to be larger than required to compensate for these losses.

In small, excavated reservoirs, seepage can be controlled by lining the impoundment with compacted clay or a synthetic liner. Evaporation can also be controlled by covering a reservoir with floating plastic sheets (Figure 3-11) or by applying a thin layer of a liquid chemical to the surface.

Siltation

All streams carry sediment. The sediment settles out when the velocity of the water drops, as it does while the water is entering an impoundment. The rate of reservoir siltation is generally a function of both the type of soil in the watershed and how well the land is protected

FIGURE 3-11 Reservoir cover installed on Garvey Reservoir at Monterey Park Metropolitan Water District of Los Angeles, Calif.

Courtesy of Burke Environmental Products, a Division of Burke Industries, Inc.

by vegetation and management practices. In some locations, siltation is so rapid that most of the capacity of a reservoir can be lost within a few years after it is constructed.

The problem of siltation can never be completely solved, but through better land use management, it can be minimized. Enforcement of regulations governing good practice in farming, logging, road construction, and other operations that disturb the land will help reduce the amount of sediment entering tributary streams.

Creation of artificial wetlands for small streams or at the upper ends of a reservoir can also help reduce siltation. Wetlands reduce the stream's velocity and thereby cause the bulk of the sediment load to settle in the wetlands before the water flows into the impoundment. Water quality in the reservoir is also improved, because many nutrients that adhere to sediment are trapped by the wetlands and are prevented from entering the reservoir.

When the capacity of a reservoir has been decreased by siltation, there are three general solutions: the silt can be removed by dredging, the reservoir can be drained and the silt excavated, or a new reservoir can be constructed. Draining and excavating are only an option when the water system has other water sources available, and constructing a new reservoir is only possible if the necessary additional land is available. A cost analysis of the available options must be made to decide how best to restore the lost reservoir capacity.

SELECTED SUPPLEMENTARY READINGS

Linsley, R.K. et al. 1992. *Water Resources Engineering.* New York: McGraw-Hill, Inc.

Manual of Individual Water Supply Systems. 1982. US Environmental Protection Agency, Office of Drinking Water. EPA-570/9-82-004. Washington, D.C.: US Government Printing Office.

Manual of Instruction for Water Treatment Plant Operators. 1989. Albany: New York State Department of Health.

Water Quality and Treatment, 6th ed. 2010. New York: McGraw-Hill and American Water Works Association (available from AWWA).

CHAPTER 4

Emergency and Alternative Water Sources

It is very important that the operation of a public water system be uninterrupted. One of the most compelling reasons to maintain continuity of service is the protection of public health. Any loss of system pressure or interruption in service is an opportunity for microbiological or chemical contamination to enter the water system. And because water is used for sanitary purposes, a lack of water for more than a few days can contribute to the spread of disease. The possibility of water contamination and the loss of the raw water source require that every water utility develop an emergency response plan.

Table 4-1 illustrates a disaster effects matrix for a typical water utility. It lists the disasters that can affect a water system and the potential magnitude of the hazards associated with each of those disasters. More information on emergency planning is contained in AWWA Manual M19, *Emergency Planning for Water Utilities.*

An emergency response plan should be developed by each utility in cooperation with neighboring utilities and with the appropriate local and state agencies. Agencies that have a stake in the emergency response plan include local police, fire, and emergency or disaster preparedness departments, as well as the corresponding groups at the county and state levels. The purpose of this chapter is not to address the entire emergency planning program but only to examine possible emergencies involving source water. Alternative water sources for nonpotable use are also discussed.

CAUSES OF SOURCE DISRUPTION

In general, catastrophes that can disrupt the operation of a water source fall into one of two categories: natural disasters, including earthquakes (Figures 4-1 and 4-2), floods, hurricanes, tornadoes, forest fires, landslides, snow and ice storms, and tsunami; and human activities, including vandalism, explosions, strikes, riots, terrorism, warfare, and water contamination caused by leaks, spills, or dumping of hazardous materials. Risk of operator error persists under any circumstances, and machines sometimes break down on their own. Disruptions can affect the delivery of water either by directly damaging components of the treatment and distribution systems or by contaminating the water itself.

Sources of Contamination

The contamination of source water as a result of a disruption may be caused by microbiological, chemical, or radiological agents. Both groundwater and surface water sources can be affected. Some relatively common examples of contamination incidents are:

- A pipeline that has broken upstream of a surface water intake, allowing several hundred gallons (liters) of fuel oil to be spilled into the river.

- A tank truck that has overturned on a bridge and has spilled a quantity of toxic chemical into the river.
- A spill of chemical solvent on the ground near a well that has penetrated to the aquifer and is now showing up in the well at increasing concentrations.
- Excessive rainfall resulting in excessive runoff entering a combined sewer; the result is a combined sewer overflow that sends untreated sewage into a nearby body of water.

TABLE 4-1 Hazard summary for a hypothetical water system

Type of Hazard	Estimated Probability	Estimated Magnitude	Comments
Earthquake	1 in 60 years	7.0 (Richter scale)	
Fault rupture	Medium	2 ft (0.6 m)	Meridian fault
Ground shaking	High		
Liquefaction	Medium–low	Vertical and horizontal accelerations	Fill areas
Densification	Medium		Fill areas
Landslide	Medium–high		In slopes of 30 percent
Tsunami and seiche	None		
Hurricane	None		
Wind			
Storm surge			
Flooding			
Tornado	Low		
Flood	Low–medium	100-year flood to elevation = 1,020 ft (311 m)	At treatment plant
Forest or brush fire	High		Dry Creek Watershed
Volcanic eruption	1 in 300 years	150 miles away (241 km)	Mount Nueces

Table continued next page

TABLE 4-1 Hazard summary for a hypothetical water system (Continued)

Type of Hazard	Estimated Probability	Estimated Magnitude	Comments
Other severe weather			
Snow or ice	None		
Extreme heat	High	100-year drought	Reservoirs depleted
Wind	Medium	60–80 mph (97–129 km/hr)	Usually in winter
Lightning	Low		
Other			
Waterborne disease	Low		Cryptosporidiosis
Hazardous-material release			
Chlorine	Medium–high	1-ton containers	Earthquake damage
Other spill	Medium	Tanker car	Dry Creek Reservoir
Structural fire	Low		
Construction accident	Medium	Line damage	In older area of system
Road	Low		
Rail	Medium		Rail yard near warehouse
Water	Low		
Air	Low		
Nuclear power plant accident	Low	Contamination	Lake West Reservoir
Nuclear bombs explosion	Low		
Vandalism, terrorism	Medium		Storage tanks
Riot	Low		
Strike	Low		

FIGURE 4-1 A 400,000-gal (1.5-million-L) steel tank uplifted during the 1992 Landers, Calif., earthquake
Source: M.J. O'Rourke, RPI, NCEER.

When contamination of a water source occurs, the first thing that must be determined is the length of time of the contamination episode. In the previous examples, the tank-truck spill will probably pass the river intake in a matter of hours. The fuel-oil spill could last for several days.

Contamination of an aquifer can create a very long-term, if not permanent, problem. Whether the contamination plume passes by naturally or is deliberately cleaned up, the process may take years. In some cases, special wells can be operated around the contaminated portion of an aquifer so that the pollutant is pulled toward those wells and does not spread further.

When source water becomes contaminated, a utility can respond in several ways to ensure a continuous and wholesome water supply: It can stop drawing water and operate on stored water until the episode is over, change to an alternative water supply until the episode is over, or treat the contaminated water.

The concentration of most chemical contaminants in water can be reduced to acceptable levels with the proper type of treatment, such as aeration or carbon adsorption. Microbiological contaminants can generally be controlled by proper coagulation, sedimentation, and filtration followed by adequate doses of disinfectant. However, the type and extent of equipment required for special treatment during a contamination episode may not be required for normal source treatment operations. Therefore, water system operators should anticipate the contamination problems that might occur and have plans

FIGURE 4-2 Flocculator/clarifier center mechanism damaged during 1989 Loma Prieta earthquake
Source: D.B. Ballantyne.

for bringing any special equipment on-line in the case of an emergency. Additional details on water treatment methods for contaminant removal are included in *Water Treatment*, also part of this series.

Notification of Possible Contamination

Utilities should coordinate with their local emergency planning department to establish specific procedures for contacting local, state, and federal officials should an incident occur. If the contamination is the result of an intentional act, utilities should also notify local law enforcement, as well as state and federal authorities (including homeland security agencies). Further information is available from the US Environmental Protection Agency's *Guidance for Water Utility Response, Recovery, and Remediation Actions for Man-made and/or Technological Emergencies.*

Loss of Water Source

A complete loss of a water source can occur under several scenarios: when an earthquake has damaged all of a water system's wells, when a flash flood on a river has destroyed a system's intake facilities, or when systems fail because of a pipeline break or tunnel collapse.

If a water system's usual water source is no longer available, the utility will need to determine quickly whether the damaged component is repairable and whether a temporary solution can be put in place. Damaged wells, for example, may not be readily repairable, in which case new wells may have to be drilled, and that could take months. But if a surface water intake is destroyed, it may be possible to install a temporary pipeline or bring in portable pumps and have them operational within a few days.

Short-term options

The following short-term options should be considered when a water source is damaged or destroyed:

- If a limited supply of water is still available, commence conservation or rationing.
- Supply customers with water from tank trucks.
- Supply customers with bottled water.
- Draw water from adjoining water systems.

Long-term options

The following long-term options should be considered when a water source is damaged or destroyed:

- Drill new wells.
- Construct a new surface water source.
- Clean up the source of contamination.
- Install a connection to draw water from another water system.
- Impose permanent conservation requirements.
- Develop additional sources for nonpotable uses.
- Reuse wastewater for nonpotable requirements.
- Install a dual potable–nonpotable water system.
- Construct new raw-water storage, such as a new reservoir.
- Perform aquifer recharge.

Evaluating the options

Emergency response options are best evaluated by carefully considering and comparing these various factors:

- Technical and logistical feasibility
    - What procedures are required to implement the option?
    - Is the required technology available and properly developed?
    - How much water can the option provide?

- Will the option meet only the system's priority water needs?
- Can it meet the system's current total water needs?
- Can it be expanded to meet future community water needs?
- How quickly can the option be made operational?
- What equipment and supplies are needed?

- Reliability
  - How reliable is the option?
  - Does it require special operation and maintenance skills?
- Political and legal considerations
  - What administrative procedures are required to implement the option?
  - Is property ownership an issue?
  - Are water rights an issue?
  - Will the option be acceptable to the public?
- Cost considerations
  - What initial investment is required?
  - What will the operating costs be?
  - Who will bear the cost of the design, construction, and operation of the option?

ALTERNATIVE WATER SOURCES

If at all possible, alternative water sources should be identified before an emergency that requires an alternative source actually happens. Additional sources of material to assist in preparing for both natural and human-caused disasters are given in the list of Selected Supplementary Readings at the end of this chapter. Some typical alternative sources that should be considered are discussed below.

- Surface water systems
  - Provide two or more intakes at different locations in the lake or stream.
  - Provide intakes in more than one water source, if possible.
  - Construct wells for backup of the surface source.
- Groundwater systems
  - Provide enough wells to meet demand when one or more wells are out of service.
  - Locate wells in different aquifers—or far enough apart in the same aquifer—so that contamination affecting one well will not affect other wells.
  - Provide a surface water emergency source.
- All systems
  - Establish interconnection with other water systems that have sufficient capacity to meet at least minimum needs.

Some emergency situations can be avoided by good practices in facility design and operation. For instance, good security at water facilities can reduce vandalism; providing standby power eliminates complete dependence on the commercial electric power grid; and locating all facilities far enough away from and well above floodplains (of both rivers and coasts) can prevent inundations and many contamination problems. In addition to averting problems that can possibly be prevented, water system operators should be prepared to act swiftly and efficiently in the event of an emergency.

INTERSYSTEM CONNECTION

In urban and suburban areas, it is not unusual for water mains belonging to different water utilities to lie close to each other in some locations. Even without prior planning, emergency connections can be made between the distribution systems using fire hoses or temporary piping. Unfortunately, this consumes precious time during an emergency and usually provides only a fraction of the water needed.

Consequently, it is often advantageous to interconnect adjoining distribution systems or raw-water transmission systems. If water mains are large enough at adjacent sites and water pressure is about the same, the interconnection can be relatively simple. A valve can be kept closed until an emergency need arises to share water between the systems.

The managers of adjoining water systems should carefully plan for either system to furnish water to the other in an emergency. Before establishing a connection, system managers should determine whether the separate water supply systems have a compatible water chemistry and quality to allow intermingling. A satisfactory connection will require piping of sufficient size to provide the needed amount of water in either direction. If system pressures are not compatible, it may be necessary to install pressure-reducing valves and a booster station. It may also be desirable to provide meters for flow in either direction, for purposes of cost reimbursement and water accountability.

EMERGENCY PROVISION OF WATER

When the distribution system is providing only a reduced amount of water, emergency supplies can be brought in by tank trucks and as bottled water. Rationing can also be used while the distribution system is operating at reduced capacity.

Tank Trucks

If a water distribution system suffers massive failure and no alternative piped source is available, nonpiped water must be used. Tank trucks (Figure 4-3) can be used for this purpose, and every utility should maintain a record of where suitable units are available. The utility must be able to confirm the chemical and biological safety of the water supplied by the trucks and must ensure that adequate disinfection is maintained.

Drivers of tank trucks must take the essential sanitary precautions and, most importantly, must bring water from a wholesome water source. If water is to be taken from a

FIGURE 4-3 Emergency potable water provided by tanker truck in Kauai, Hawaii
Source: Ray Sato.

neighboring utility, communication and monitoring are necessary to ensure that the water-filling point is satisfactory to both utilities.

Bottled Water

Bottled water is also frequently made available to customers during an emergency. Because demand can quickly exceed the normal supply of bottled water in an area, it may be necessary to tap into supplemental supplies of bottled water. In some cases, local dairies and beverage bottlers can provide water in their own milk cartons or bottles during an emergency. Beverage companies may also have trucks available that can be used for delivery. In any case, the cost of bottled water will be very high compared to the cost of water from the public distribution system.

Water Rationing

Some disasters reduce the supply of water, even when the distribution system is still functioning. Typical situations include the following:

- The system can reestablish the water source soon and has enough water in storage to supply basic customer needs if use is drastically reduced.
- An interconnection has been established with another utility, but it provides only enough water to supply basic needs.

- The water system must operate at very low pressure because of pump failure or other problem. Care should be taken if the system operates at a low water pressure because the risk of contamination is higher due to backfill or groundwater infiltration.

In these cases, a drastic reduction in demand can help alleviate the crisis. The utility's conservation plan or emergency response plan should have all categories of water use rated by priority. This situation is comparable to that faced by communities with severe or critical drought conditions. Table 4-2 is an example of a plan prepared for use in such situations.

WATER REUSE

As populations grow and potable water becomes scarcer in some areas, the incentive for reusing water increases. Water may be used indirectly, directly, and through the provision of dual water systems.

Indirect Reuse

Indirect potable reuse—the use of water from streams that have upstream discharges of wastewater—has been done for a very long time. Discharge from individual septic systems infiltrates into aquifers and is filtered by the soil and aquifer formation, and the water is later withdrawn through wells. Sewage and industrial wastes discharged to rivers are diluted and somewhat treated by natural processes (such as photoreaction, oxidation, and decomposition), and the water is later withdrawn by downstream communities for their water supply. Every day millions of people worldwide consume water that has been used somewhere upstream.

Indirect reuse will certainly increase with urban population growth and land development as reuse becomes necessary along streams in a watershed. This trend of population increase and land development does threaten the safety of water supplies; however, as long as both the dischargers and receivers are aware of the factors influencing water quality and treat both the wastewater and the potable water to meet appropriate instream water quality standards and drinking water standards, the public health risk is minimized.Recharging of groundwater through surface spreading or direct injection of reuse water is being practiced on a large scale in some areas. The groundwater can then be pumped for nonpotable uses (such as irrigation), and demand on the potable water aquifer is reduced.

Direct Reuse

The direct reuse of wastewater for purposes not requiring potable water can help meet a significant portion of the water demand in areas where water is particularly scarce. Wastewater may include treated municipal sewage, industrial cooling and process water, and agricultural irrigation runoff water. Any water that is used again after being used by any consumer is classified as reclaimed water. Currently, the principal use of reclaimed water is for agricultural irrigation. Figure 4-4 shows a typical irrigation concept using reclaimed water. For decades, certain areas of the United States, such as the southwestern states, have

TABLE 4-2 Example of a drought or emergency conservation plan

Stage	Degree of Drought or Emergency	Consumption Reduction Goal, %	Public Information Action	Public-Sector Action	User Restrictions	Penalties for Noncompliance
I	Minor	10	Explain drought or emergency conditions. Disseminate technical information. Explain other stages and possible actions. Distribute retrofit kits at central depots. Request voluntary reduction.	Increase enforcement of hydrant-opening regulations. Increase meter-reading efficiency and meter maintenance. Implement intensive leak detection and repair program.	Voluntary installation of retrofit kits. Restriction of outside water use for landscape irrigation, washing cars, and other uses.	Warning
II	Moderate	15–18	Use media intensively to explain emergency. Explain restrictions and penalties. Explain actions in potential later stages. Request voluntary reduction.	Reduce water usage for main flushing, street flushing, public fountains, and park irrigation.	Mandatory restriction on all outside uses by residential users, except landscape irrigation. Prohibition of unnecessary outside uses by any commercial users.	1. Warning 2. House call 3. Installation of flow restrictor 4. Shutoff and reconnection fee
III	Severe	25–30	Public officials appeal for water use reduction. Explain actions and consequences of emergency.	Prohibit all public water uses not required for health or safety.	Severely restrict all outside water use. Prohibit serving water in restaurants. Prohibit use of water-cooled air conditioners without recirculation.	Same as Stage II
IV	Critical	50 or more	Same as Stage III	Reduce system pressure to minimum permissible levels. Close public water-using activities not required for health or safety.	Prohibit all outside water use and selected commercial and industrial uses. Terminate service to selected portions of system as last extreme measure.	Same as Stage II

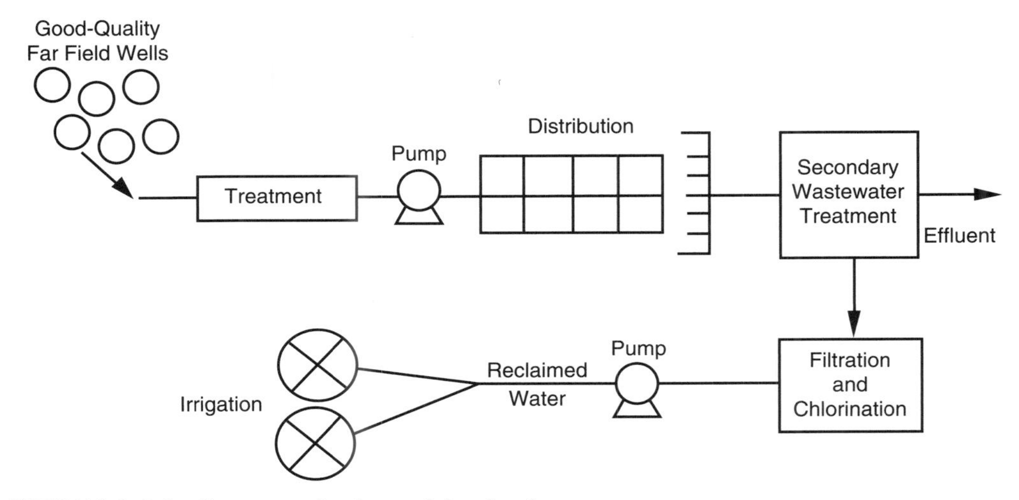

FIGURE 4-4 Irrigation concept using reclaimed water

been successfully applying reclaimed water to golf courses, orchards, and nonedible crops and using it for flushing toilets (Figure 4-5).

In 2004, AWWA adopted this policy concerning Reclaimed Water for Public Water Supply Purposes:

> The American Water Works Association encourages responsible use of reclaimed water instead of potable water for irrigation, industrial, and other non-potable uses within a public drinking water supplier's service area when such use can reduce the demands placed on limited supplies of potable water.
>
> In cases where raw water supply sources are limited, AWWA recognizes the value of indirect use of reclaimed water to supplement existing raw water sources. These waters must receive appropriate subsequent treatment and be acceptable to health authorities and water users.
>
> AWWA encourages continued research to improve treatment technology, monitoring techniques, and the development of health-based drinking water standards that assure the safe use of reclaimed water.

Dual Water Systems

One method of achieving water reuse is to deliver potable water and nonpotable water to consumers through separate distribution systems. This is generally referred to as using a

FIGURE 4-5 El Tovar Lodge in Grand Canyon Village, Ariz., the site of the first US dual distribution system in the United States (reclaimed water is used for landscaping and toilet flushing)

dual water system. Use of a dual system is also a possible solution where potable water is in very limited supply but lower-quality nonpotable water is more plentiful. The concept is not new and has been used occasionally by water systems having unusual source water conditions.

Analyzing water demand, especially domestic use, gives a better picture of the potential for the great savings of potable water through the use of dual water systems. In the home, high-quality potable water is needed for drinking, cooking, bathing, laundry, and dishwashing. The other uses, primarily toilet flushing, do not necessarily require high-quality water.

Exterior water use as a percentage of total use varies dramatically with the seasons and geographic location. On an annual basis, exterior use averages 7 percent in Pennsylvania and 44 percent in California. Because high-quality water is not required for exterior residential uses, the demand can be satisfied by the use of nonpotable water.

Dual water systems are in use in several areas of the United States, including Tucson and Phoenix, Arizona; Irvine, California; and Colorado Springs and Denver, Colorado.

Probably the most extensive dual water system is in St. Petersburg, Florida.; Figure 4-6 shows its water reclamation plant. The local water supply there was exhausted in 1928 because of saltwater intrusion into the water system's wells. Since that time, other well fields have been developed, some as far as 50 mi (80 km) away. A great portion of the water use in the city is for irrigation of domestic lawns, as well as municipal and institu-

FIGURE 4-6 Reclamation plant in St. Petersburg, Fla., that incorporates secondary treatment, filtration, and chlorination
Source: St. Petersburg Department of Public Utilities.

tional lawns and green spaces. A very detailed financial study was undertaken comparing the cost of various alternatives. Overall savings after installing a dual water system include the savings from not having to develop more potable-water source capacity. An added benefit arising from the nature of the dual system is the savings on fertilizer costs because of the amount of nitrogen, phosphorus, and potassium in the reclaimed water.

It is likely that more dual systems will be constructed in the future, but public health consideration is paramount in their design: Improper use of nonpotable water must be controlled, cross-connections strictly monitored, and responsibility for the nonpotable system clearly established.

SELECTED SUPPLEMENTARY READINGS

Drinking Water System, Emergency Response Guidebook. 2001. Salt Lake City, Utah: Department of Environmental Quality, Division of Drinking Water.

Guidance for Water Utility Response, Recovery and Remediation Actions for Man-made and/or Technological Emergencies. 2002. EPA 810-R-02-001. Washington, D.C.: US Environmental Protection Agency, Office of Water, Water Protection Task Force.

Guide to Ground-Water Supply Contingency Planning for Local and State Governments. 1991. Washington, D.C.: US Environmental Protection Agency, Office of Drinking Water.

Manual M19, Emergency Planning for Water Utility Management. 2001. Denver, Colo.: American Water Works Association.

Manual M24, Planning for the Distribution of Reclaimed Water. 2009. Denver, Colo.: American Water Works Association.

Minimizing Earthquake Damage: A Guide for Water Utilities. 1994. Denver, Colo.: American Water Works Association.

Security Analysis and Response for Water Utilities. 2001. Denver, Colo.: American Water Works Association.

States, Stanley. 2010. *Security and Emergency Planning for Water and Wastewater Utilities.* Denver, Colo.: American Water Works Association.

Water System Security: A Field Guide. 2002. Denver, Colo.: American Water Works Association.

CHAPTER 5

Use and Conservation of Water

While water is one of our most vital natural resources, the availability of clean water at the tap has often been taken for granted. But recent shortages, quality problems, and cost increases in various parts of the country have resulted in a new appreciation for the challenge of providing a continuous and wholesome water supply.

A substantial part of the cost of water is due to the expense of treatment and distribution. Additional costs are incurred by utilities to comply with the Safe Drinking Water Act (SDWA) and other environmental regulations. Drinking water is being treated and tested more than ever. Increased health-related concerns and ongoing advances in commercially available technology continue to drive improvements in the effectiveness of treatment and monitoring performed by water utilities, resulting in increased capital and operating costs.

It is worth emphasizing, however, that the value of water as a resource cannot be measured simply by the costs associated with water treatment and distribution or the price paid by customers. Wise management of water recognizes not only the cost of obtaining it in a useful form, but also the value of the benefits it delivers—not only in society, but in the natural world.

WATER USE

Public water supply uses can be grouped into three categories: domestic, industrial, and public.

Domestic Use

The category of domestic water use includes water that is supplied to residential areas, commercial districts, and institutional facilities.

- **Residential uses.** Water provided to residential households serves both interior uses (toilets, showers, clothes washers, faucets, etc.) and exterior uses (car washing, lawn watering, etc.). The amount of water used by households varies according to such factors as the number and ages of occupants in a household, income level, geographic location (which influences climate), and the efficiency of water fixtures and appliances in the home. Typical data for rates of interior water use are presented in Figure 5-1. Outdoor water use is strongly influenced by the annual weather patterns; water use in hot, dry climates is a significantly higher than in colder, wetter climates. Table 5-1 shows a typical breakdown of water use by a typical family of four persons living in a single-family house.

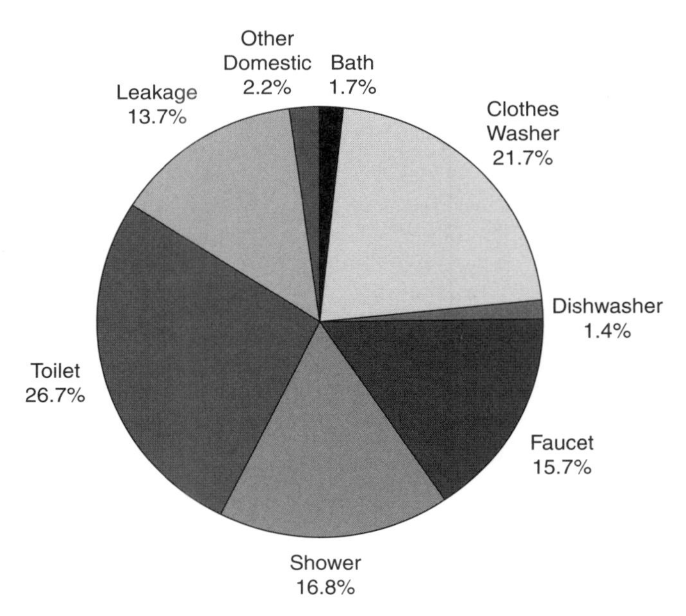

FIGURE 5-1 Indoor per capita use percentage by fixture

- **Commercial.** Water use can vary widely among commercial facilities depending on the type of commercial activity. A small office building would be expected to use less water than a restaurant of the same size, for example. In addition to the amount of water required by employees and customers for drinking and sanitary purposes, commercial users may need water for such diverse uses as cooling, humidifying, washing, ice making, vegetable and produce watering, ornamental fountains, and landscape irrigation.
- **Institutional.** Estimates of water demand for facilities such as hospitals and schools are usually based on the size of the facility and the amount of water needed for each user in the facility. Water use by schools is highly dependent on whether the students are housed on campus or are day students.

Industrial Use

In the industrial sector, rates of water use depend on the type of industry, cost of water, wastewater disposal practices, types of processes and equipment, and water conservation and reuse practices. In general, industries that use large quantities of water are located in communities where water quality is good and the cost is reasonable.

Historically, major industrial water users (such as the steel, petroleum products, pulp and paper, and power industries) have provided their own water supply. Smaller industries and industries with low water usage are more inclined to purchase their water from public systems.

TABLE 5-1 Typical residential end uses of water by a family of four*

Fixture/End Use	Average gallons per capita per day	Average liters per capita per day	Indoor use, %	Total use, %
Toilet	18.5	70	30.9	10.8
Clothes washer	15	56.8	25.1	8.7
Shower	11.6	43.9	19.4	6.8
Faucet	10.9	41.3	18.2	6.3
Other domestic	1.6	6.1	2.7	0.9
Bathtub	1.2	4.5	2.0	0.7
Dishwasher	1	3.8	1.7	0.6
Indoor Total	59.8	226.4	100.0	34.8
Leakage	9.5	36	NA	5.5
Unknown	1.7	6.4	NA	1.0
Outdoor	100.8	381.5	NA	58.7
TOTAL	171.8	650.3	NA	100.0

*Data collected from the following cities: Boulder, Colo.; Denver, Colo.; Eugene, Ore.; Seattle, Wash.; San Diego, Calif.; Tampa, Fla.; Phoenix, Ariz.; Tempe, Ariz.; Scottsdale, Ariz.; Cambridge, Ont., Canada; Waterloo, Ont., Canada; Las Virgenes, Calif.; Walnut Valley, Calif.; and Lompoc, Calif. (Adapted from AwwaRF 1999).

Public Use

Municipalities and other public entities provide public services that require varying amounts of water. Water is used for

- public parks, golf courses, swimming pools, and other recreational areas,
- municipal buildings,
- firefighting, and
- public-works uses such as street cleaning, sewer flushing, and water system flushing.

Although the total annual volume of water required for firefighting is typically small, the rate of flow required during a fire can be very large. The flow rate needed to fight a large fire in a small community can put a significant strain on the water system and may threaten to reduce pressure in the system substantially. For example, a community that uses an average of 1 mgd (4 ML/d) or 694 gpm (2,637 L/min.) might reasonably expect a fire flow demand of 3,000 gpm (11,400 L/min.) for up to 3 hours. The rate of flow for firefighting in this case is more than four times the average daily flow, and it will likely be required from a few water mains near the fire.

Variations in Water Use

Water use for a municipality or community can vary due to several factors, including:

- Time of day and day of the week
- Climate and season of the year
- Type of community (residential or industrial) and the economy of the area
- Presence or absence of customer meters
- Dependability of supply and dependably high quality of the water
- Condition of the water system (leakage and losses)
- Water conservation/demand management

Time of day and day of week

Water use rates vary considerably according to the time of day and the day of the week. On a typical day in most communities, water use is lowest at night (11 p.m. to 5 a.m.) when most people are asleep. Water use rises rapidly in the morning (5 a.m. to 11 a.m.), and usage is moderate usage through midday (11 a.m. to 6 p.m.). Usage then increases in the evening (6 p.m. to 10 p.m.) and drops rather quickly around 10:00 p.m. A diurnal curve showing the hourly rate of water use for a community on a typical day is shown in Figure 5-2.

Total water use rates for a community typically vary by the day of the week as well, depending on the habits of the community. Some water systems have significant increases in water use on Monday because in many households Monday is washday. Water systems supplying industries that do not operate on weekends observe significantly reduced usage rates on Saturday and Sunday.

Climate and season

Water use is typically highest during the summer months when the weather is hot and dry and the need to provide water for exterior activities and landscape irrigation is at its peak. Particularly significant seasonal variations are commonly observed in resort areas, small communities with college campuses, and communities with seasonal commercial or industrial activities.

Most systems have relatively low water use in winter when there is no need for lawn or garden sprinkling and there is less outside activity requiring water. On the other hand, in extremely cold weather consumers may occasionally run water faucets continuously to prevent water pipes from freezing.

Air-conditioning units and cooling towers operate by evaporating water and are used extensively by homes and large commercial and industrial facilities located in hot, dry climates. The demand for water to support air conditioners and cooling towers generally coincides with periods of highest water demand for lawn and garden sprinkling, pushing the annual water use to its peak for the year.

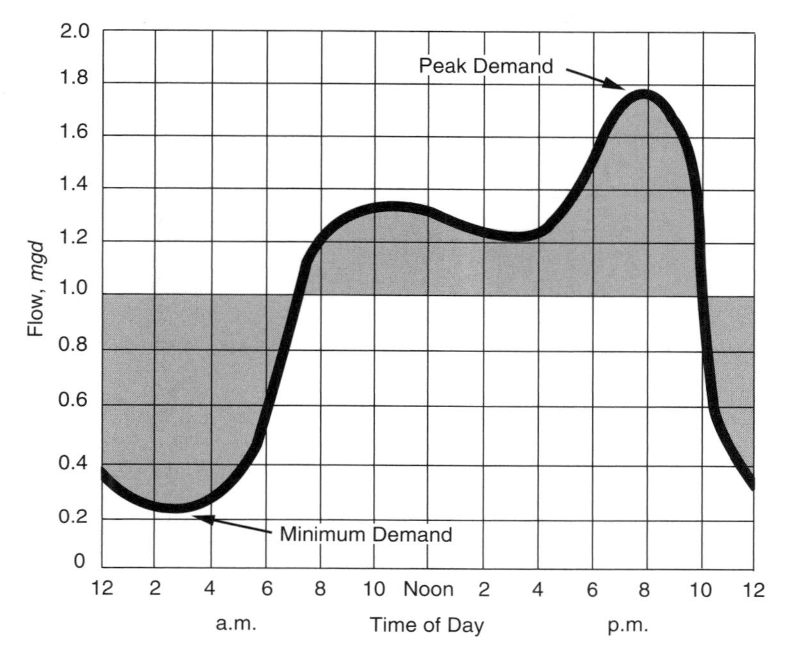

FIGURE 5-2 Typical daily flow chart showing peak and minimum demands

Type and size of community

The type of housing in the service area will affect rates of water use. Areas that have individual homes on large lots with gardens and lawns typically have a much higher water use per person than areas with multiple-family dwellings such as townhomes, condominiums, and apartment complexes. Households with automatic washing machines, dishwashers, and garbage disposals tend to use more water than households where these appliances are not installed. Economically depressed areas generally have lower water use per capita.

Community size typically affects the variability of water use rates over time. The ratio of peak use to average use tends to be greater for systems serving small communities than for larger water systems.

Metering

Most water utilities install meters that track all water usage. Other systems do not have customer meters and continue to charge customers a flat rate regardless of the quantity of water they use. The choice of billing scheme significantly influences the amounts of water consumed. Flat-rate customers may use as much as 25 percent more water than is used by customers who have to pay more as they use more. This excess water usage can be caused by unnecessary and unrestricted sprinkling and by personal behavior (e.g., letting a faucet run to get cold water or letting a hose run while washing a car). Another consequence of flat-rate billing is that household leaks are often ignored because wasted water costs the customer nothing.

Dependability and quality of water

Customers generally use less water from the public system if the water pressure is poor or if the water has an unpleasant taste, odor, or color. They may also use less if it has a high mineral content. Under such conditions, customers may choose alternative water sources, some of which may be less protective of public health. Bottled water, for example, has been subject to less stringent regulatory standards than tap water. (General Accounting Office, 2009) Private wells, likewise, are not as closely regulated as a public utility.

Sewer connection

The availability of municipal sewer systems also increases water use. Homes with sewer service usually have more water-using appliances and fixtures because there is no concern about overloading an on-site wastewater treatment and disposal system that would be used otherwise. The increase in water use due to sewer availability may be 50 to 100 percent.

Condition of the water system

The difference between the average annual water production and the average annual consumption of a system amounts to ongoing unaccounted-for losses. Unaccounted-for losses can result from many factors, including leaks in piping, main breaks, periodic fire hydrant flushing, tank drainage for maintenance purposes, unauthorized use, unmetered services, and inaccurate or nonfunctioning meters. Current industry standard states that unaccounted-for system leakage and losses should be less than 10 percent of production for new distribution systems (those less than 25 years old) and less than 15 percent for older systems. Rates of leakage are affected by the age of the system, the materials used in construction, and the level of system maintenance. Water mains are usually tested for watertightness when they are installed. However, as the water mains get older, leakage at pipe joints, valves, and connections increases. Unless the system is regularly checked for leaks and unless leaks are promptly repaired, the amount of water lost daily can be significant. Such losses not only place an extra strain on the ability of the utility to furnish water to customers, they also represent an important loss of revenue.

WATER CONSERVATION

Definitions of water conservation vary. In this book, water conservation is considered to be any beneficial reduction in water use or in water losses. Over the past four decades, the level of appreciation for water conservation efforts has increased significantly. Water utilities have been among the leaders in conservation efforts. For utilities in areas of prolonged drought, the benefits of reducing water use are obvious, and mandatory restrictions may be put into effect where the drought is severe. The benefits are less evident in an area that has adequate rainfall, a relatively stable population, and a water source with an adequate safe yield. A useful approach to evaluating whether a conservation program is worthwhile is to compare the expected benefits against the anticipated costs.

Potential Benefits of Water Conservation

Some potential benefits of water conservation are:

- **Reduced per capita demand on the supply source.** A reduction in water use will allow a larger population to be served by the water source. This is particularly important where aquifers are being depleted and limits on the amounts of available surface water are being approached by growing communities. Reducing normal rates of water use can also help ensure that adequate supply is available during an emergency.
- **Reduced need for facility capacity.** Reduced flows in the water and wastewater system may allow a water utility to eliminate or downsize new facilities and defer the construction of new ones.
- **Reduced costs for utility and customer.** A reduction in water use can provide energy savings for the water utility. Pumping costs can be reduced, as can the costs for power and chemicals required for all stages of treatment. Cost savings can help defer increases in billing rates for consumers. In addition, when heated water is conserved, energy costs will be reduced in homes, businesses, and industry.
- **Protection of the environment.** Reducing the rate of withdrawals from a stream, lake, or impoundment can help maintain flow rates and water levels needed to sustain a healthy environment for native aquatic plants and animals.

Potential Problems With Water Conservation

Some undesirable consequences of implementing a water conservation program are:

- **Loss of revenue for the utility.** When a utility implements a water conservation program and does not increase water rates, it will receive less income because of reduced water sales. There will be some offsetting savings to the utility due to lower energy and chemical costs, but the net loss can still be substantial. Often the only way to compensate for this loss in revenue is to raise water rates. Before starting a conservation program, the water system must plan for a decrease in revenue and take steps to ensure that all financial obligations, including bond payments, will still be met. Because it is usually difficult to raise rates on short notice, advance financial planning is particularly important should it ever become necessary to implement an emergency conservation effort during a water shortage.
- **Utility and customer program costs.** The cost for implementing a conservation program will likely be borne by the water utility and the water utility's customers. Costs

for programs can vary depending on the implementation method used, the size of the program, and the scope of program evaluation (i.e., surveys, billing data reviews, water audits, etc.). To the extent possible, these short-term and long-term costs should be estimated ahead of time for each component of the water conservation program.

- **Possible delay in developing additional source capacity.** When a water crisis is alleviated through conservation, the public may find it hard to understand why there is still a need for long-range water resource development projects. Lack of public support could then delay (or eliminate from further consideration) development of a new reservoir, well field, or other new water source that is still needed. This delay can, in turn, increase the cost of the project, as well as worsen the water crisis if there is a drought year before the new facilities can be constructed.
- **Possible stimulation of water service growth**. The water saved by a conservation program will allow more people to be served by the amount of water available. This can act as a stimulant for growth. As long as water still appears to be available for use, additional customers may continue to be connected to the system. This action will not be in the best interests of the water system or community, at least until a more ample supply can be tapped, if it increases the risk that shortages will occur.
- **Difficulty in dealing with drought conditions**. When a drought occurs after a utility has a conservation program in effect, the situation can become very difficult. It is sometimes hard to explain why it is necessary to conserve even further. This situation emphasizes the need to update water demand forecasts and continue overall water supply planning beyond any conservation effort. The utility must convince customers that looking to the future water supply needs of a community is as important as any aspect of a conservation program.

Supply Management Techniques

In conjunction with any water conservation behavior expected from the public, a utility should examine its own practices in properly managing the available supply of water. This process is called supply management. Some features of supply management are:

- **Careful management of all resources.** Maintaining an adequate source of supply should be part of both long-range (25–50 years or more) and short-range (5–10 years) plans. The plans should be in writing and circulated to all local interest groups for comment and review.
- **Analysis of water use data.** A water use database should be available to allow evaluation of records and forecasts concerning average daily production, peak demand, and geographical and seasonal water use patterns. The complexity of this database should grow with utility size so that any unexpected change affecting the utility can be analyzed for its impact on total water use.

- **Complete source and customer metering.** All sources of supply and all customer services should be metered. Complete metering is necessary so that the amount of water used by customers can be compared against the amount pumped to the system.
- **Reduction of unaccounted-for water.** The difference between the amount of water pumped to the system and the amount metered to customers (or otherwise accounted for) is called unaccounted-for water. Unaccounted-for water is generally water that is wasted through leaks or unauthorized use. State utility commissions usually have formulas that can be used to calculate unaccounted-for water. Unaccounted-for system leakage and losses may range from 5–10 percent of production for new distribution systems (less than 25 years old) and up to 15 percent for older systems. If the amount of unaccounted-for water exceeds 15 percent, a systemwide leak survey should be conducted. In addition, leak awareness by field personnel can be promoted by a program of listening for leaks at hydrants, services, and meters.

Demand Management Techniques

In conjunction with the program to manage the available supply, a utility must develop and implement demand management programs to encourage efficient water use. Demand management can consist of these components:

- Public education
- Distribution of water-saving devices
- Changes in consumer behavior
- Modification of the water rate structure
- Water-saving plumbing fixtures
- Landscaping and agricultural practices that require less water
- Promotion of water conservation by businesses and industries

Public education

Educating the public is usually a key component of a water conservation program. If water rates are already low or moderate, customers may need some incentive to participate in a water conservation program. Printed materials and suggested programs are available from the American Water Works Association (AWWA), state water organizations, state agencies, and the US Environmental Protection Agency (USEPA), as well as environmental and conservation organizations. An effective way of reaching and informing the public about a conservation program is by offering speakers, literature, and videos to local schools, service clubs, and civic groups.

Customers can also be educated directly by bill stuffers and conservation notices. Direct assistance in detecting leaks or developing water use audits can also be offered to customers. Newspapers and radio and television stations can be supplied with public service announcements (PSAs) and press releases explaining the program. A utility's web site can be a very effective channel for explaining the program to the public.

Water-saving devices

Many water utilities offer water-saving kits to residential customers. Installation of these kits will decrease the flow of water through several household plumbing fixtures. A typical kit includes faucet aerators, a low-flow showerhead or flow restrictor, and a toilet flapper to reduce leakage. Some utilities offer rebates to customers who replace older-model toilets with low-flow models. The cost of these kits can be fully paid by the customer, partially paid by the customer, or fully funded by the utility. Some state regulatory agencies have mandated the use of these kits, stipulating that the cost be shared between the customer and utility.

Changes in consumer behavior

Some branches of a water system may experience water shortage as a result of inadequate distribution system capacity. If this occurs, the utility may gain some relief by educating the public about the problem and encouraging customers to use water during off-peak hours. Water use restrictions can be either mandatory or voluntary. Customers may be advised to save water by watering lawns at night, when less water evaporates, for example. Mandating that lawn watering or car washing be done only on alternating days is a common summer water use restriction.

Rate structure modification

Water can be priced in a way that encourages efficiency and discourages excessive and wasteful use. Information on various approaches to changing the rate structure for water can be obtained from AWWA and public utility regulatory agencies.

Water-saving plumbing fixtures

The Energy Policy Act of 1992 established water conservation standards for four types of plumbing fixtures: toilets, kitchen and lavatory faucets, showerheads, and urinals. With limited exceptions, the standards apply to all models of the fixtures manufactured after Jan. 1, 1994. Approximately 16 states and six localities had water-efficient standards for at least two of the plumbing fixtures regulated under the Energy Policy Act before the national standards took effect in 1994. Studies have shown that, compared to their less efficient counterparts, low-flow fixtures conserve water, particularly in the case of toilets.

Landscaping and agricultural practices that require less water

Landscaping that requires little water is called Xeriscaping (Figure 5-3). It is being encouraged for homes and businesses in semiarid and arid areas. Excellent literature is available on this subject (see Suggested Supplementary Readings at the end of this chapter). The use of drought-tolerant landscaping also has value in areas where there are seasonal shortages of water.

Although agricultural operations do not require treated water from a municipal water supply system, a municipality and the surrounding agricultural community may both draw raw water from the same source. Therefore, a municipal water system does have a stake in encouraging the most efficient use of water for agricultural purposes. Greater water use efficiencies in agriculture may be obtained by changes in tillage practices, irrigation scheduling, and technology choices such as drip irrigation and by cultivation of crops that require less water.

FIGURE 5-3 Typical water-efficient landscaping
Source: Doug Hanford, Hanford Company Landscape Design; John Nelson, North Marin Water District; Ali Davidson, Sonoma County Water Agency.

Business and industry conservation

Utilities can often achieve significant reductions in water use by providing conservation incentives for business and industry. Increasing water rates is one very strong incentive. Careful evaluation of how water is used, reused, or wasted in a manufacturing plant can also lead to significant reductions. In some instances, installing smaller, more sensitive meters in businesses and industries can generate considerable revenue by more accurately recording low flows and can serve as a stimulus to water conservation. Some industries and businesses can substantially reduce their water use by installing closed-loop processes, which keep filtering and recycling the water within their facilities.

Droughts

One of the most serious threats to the ability of a water utility to meet the demands of its customers is a drought. To respond to a drought emergency, the utility will have to tap into a supplemental supply, reduce demand, or both.

Every water system should have an emergency response plan to address how water will be provided under all conceivable emergencies. Although a drought can be very serious, an immediate response is not necessarily needed. It can take weeks or months for a drought to develop to the point at which water levels in streams and reservoirs fall too low to make them useful. That time can be used to organize resources to respond before the crisis becomes acute.

Emergency procedures to reduce water use may require limited behavioral change by consumers at first, but several significant water use restrictions may need to be fully enforced if the situation becomes critical. For instance, some communities establish mandatory restrictions and issue warnings and fines for unauthorized uses such as washing a car or watering the lawn during an acute water shortage.

WATER RIGHTS

Conflicts over who owns a source of water and who gets to use it may arise among individuals, among communities, among states, between states and the federal government, and among countries.

Allocation of Surface Water

In the United States, the legal reasoning used for allocating water from surface lakes and streams follows one of two basic lines of thinking:

- **Riparian doctrine** is commonly used in the water-rich eastern part of the country, and
- **Prior appropriation doctrine** is used in the more arid western states.

Riparian doctrine

Riparian doctrine, which is commonly used in the water-rich eastern part of the country, is sometimes called the "rule of reasonable sharing." All states that follow this doctrine subscribe to the basic theory that all property owners who have land abutting a body of water have an equal right to that water. Among the states that follow this doctrine, however, there are substantial differences in how the law is applied.

Under riparian doctrine, each riparian property owner can use as much water as needed for any reasonable purpose, as long as this use does not interfere with the reasonable use by other riparian property owners.

Consequently, when a water shortage occurs, all riparian property owners must accept a reduction in supply. The amount of water allowed for any one property owned is not affected by how long the land has been owned or by how much water the owner has used in the past.

Riparian rights are a common-law doctrine. This means that they are a part of a large body of civil law that, for the most part, was made by judges in court decisions in individual cases. Statutory law, on the other hand, is enacted by a legislative body.

Because the possession of rights arises from ownership of land abutting a water body, all of the owners are treated as equals by the courts. Conflicts are then usually resolved by the court, which determines what reasonable use is. Among the factors considered in the court's decision are

- purpose of the use,
- suitability of a use with respect to the water body in question,
- the economic and social benefits of the use,
- the amount of harm caused by the use,
- the amount of harm avoided by changing any one party's use, and
- the protection of existing values.

The growth of water use in riparian-doctrine states has led to more frequent conflicts over water sources. Several of these states have tried to remedy weaknesses in the doctrine by enacting statutes requiring that permits be obtained from a state agency. These permits vary widely in the restrictions they apply and the requirements they place on the riparian owner.

The issuance of permits moves a state away from basic riparian doctrine and closer to prior appropriation doctrine. The permit systems are severely criticized by some and considered simply unnecessary by others. They have not undergone serious challenges in the courts, so their eventual success is not guaranteed. The trend toward permit systems, however, is continuing.

Prior appropriation doctrine

The appropriation doctrine concept started with the forty-niners in the goldfields of California. When there was too little water for all the mining operations, the principle applied was "first in time, first in right." The principle was given legal recognition by the courts and later made into law by western state legislatures.

The appropriation doctrine rests on two basic principles: priority in time and beneficial use.

Priority in time. When stream flow is less than what is demanded by all water users along the stream, use is prioritized based on who has been using the water for the longest period of time. Water withdrawals that began more recently must be discontinued so that withdrawals begun earlier can continue. Unlike the riparian doctrine, the appropriation doctrine does not treat riparian owners equally.

Beneficial use. While appropriation doctrine recognizes no water right based on land ownership alone, a user who has been appropriated water is entitled to it only when it can be used beneficially. Waste is therefore theoretically prohibited, and any available water beyond the amount that can be used by one appropriator is available to others. Furthermore, nonuse of the water for a long period may result in loss of the water right by forfeiture or abandonment.

Defining beneficial use remains difficult, and conflicts arise as a result. Beneficial use is determined based on two broad characteristics: the type of use (such as irrigation, mining, or municipal use) and its efficiency. Most types of water uses have been judged to be beneficial, and the majority of court cases have been concerned with whether a particular water use is inefficient or wasteful.

Legal complications. Legal solutions to water conflicts can get quite complex because water, unlike property, is a shared resource, causing the exercise of one user's water rights to affect someone else. Boundaries are not obvious and may be constantly changing.

With the appropriation doctrine, the assignment (or not) of a right at any particular time depends on the observed streamflow, the number of prior existing rights, and the amount of water needed by those who have the prior rights. For many western streams, the exercise of rights might have to be adjusted as often as daily in times of heavy use, such as in summer when irrigation demands are high.

Allocation of Groundwater

The four water rights systems used for groundwater are:

1. Absolute ownership
2. Reasonable use
3. Correlative rights
4. Appropriation-permit systems

Absolute ownership

The absolute ownership system is based on the principle that the owner of the land owns everything beneath that land, all the way down to the center of the earth. The dynamic nature of groundwater movement makes it impossible to apply this ownership literally. For all practical purposes, the system is a rule of capture. That is, the landowner can use all the water that can be captured from beneath the owner's land.

There are almost no restrictions involved with this rule. The water can be used for the owner's purposes both on and off the land, and it can be sold to others. There is no liability if pumping reduces a neighbor's supply or even dries up the neighbor's well. Some interpretations of absolute ownership allow that water can be wasted as long as it is not done maliciously.

Reasonable use

As with the rule of absolute ownership, the rule of reasonable use holds that a property owner has the right to pump and use groundwater lying beneath the property. However, if this use interferes with a neighbor's use, it may continue only if it is reasonable. Furthermore, an owner is liable if unreasonable use causes harm to others. This amounts to a *qualified right* rather than an *absolute right* to use groundwater.

The determination of what is reasonable is the important aspect of making reasonable use legally workable. Reasonable use for groundwater is much easier to determine than reasonable use for surface water under the riparian doctrine. In general, any use of water that is not wasteful for a purpose associated with the land from which the water is drawn (an overlying use) is reasonable. Conversely, any use off the property may be considered unreasonable if it interferes with the use of the groundwater sources by others.

The overlying use criterion, however, does not resolve disputes if all parties to the dispute are using the water for overlying purposes. In this case, all uses are reasonable, all can continue, and reasonable use becomes the rule of capture—in other words, the owner with the biggest pump gets the water. On the other hand, because a nonoverlying use is considered unreasonable if it interferes with an overlying use, such uses are forbidden no matter how beneficial.

Correlative rights

The rule of correlative rights holds that the overlying use rule is not absolute but is related to the rights of other overlying users. This rule is used when there is not enough water to satisfy all overlying uses. In this case, the rule requires sharing. This rule has been applied by allocating rights among property owners in proportion to the size of the overlying land parcel.

Appropriation–permit system

The appropriation–permit system is sometimes called a groundwater appropriation system. The most important aspect of it is the rule of priority: Water rights are based on who has used the water for the longest period of time. Priority is all-important in the appropriation doctrine as applied to surface water because the variation in streamflow requires frequent changes in allowable withdrawals. However, the flow of groundwater in an aquifer fluctuates very slowly under natural conditions, so that frequent adjustment in allocations is not necessary. This means that priority, as applied to groundwater, mainly involves limiting the number of permits to prevent overuse of the aquifer.

The permit system thus amounts to groundwater management and administrative regulation. By placing limitations on pumping rates, well field placement, and well construction standards and by refusing permits when necessary, for example, the administrating agency manages the groundwater in its jurisdiction. The implementation of permit systems varies widely, however. The only generalization that can be made is that permits can be more relaxed if adequate water is available and more stringent if it is not.

Conclusion

The administration of water rights can vary considerably across the United States. Consequently, a water utility operator should be familiar with the regulations that apply to the specific area that the utility serves.

SELECTED SUPPLEMENTARY READINGS

Atassi, Amrou, J. Pingatore, J. Hardwick, E. Opitz. 2009. Proactive Planning Preserves Resources for Future Generations. *Opflow* 35(12):20–21.

Bowen, P.T., J.F. Harp, J.W. Baxter, and R.D. Shull. 1993. *Residential Water Use Patterns.* Denver, Colo.: Awwa Research Foundation and American Water Works Association.

Driscoll, F.G. 1986. *Groundwater and Wells.* St. Paul, Minn.: Johnson Filtration Systems, Inc.

Drought Management Handbook. 2002. Denver, Colo.: American Water Works Association.

General Accounting Office, 2009. *Bottled Water: FDA Safety and Consumer Protections Are Often Less Stringent Than Comparable EPA Protections for Tap Water*, GAO-09-610.

Green, Deborah. 2010. *Water Conservation for Small- and Medium-Sized Utilities.* Denver, Colo.: American Water Works Association.

Hildebrand, Mark; Sanjay Gaur; Kelly J. Salt. 2009. Water Conservation Made Legal: Water Budgets and California Law *Journ. AWWA*, 101(4):85–89.

Manual M50, *Water Resources Planning*. 2007. Denver, Colo.: American Water Works Association.

Manual M52, *Water Conservation Programs—A Planning Manual.* 2006. Denver, Colo.: American Water Works Association.

Mayer, P.W. et al. 1999. *Residential End Uses of Water.* Denver, Colo.: Awwa Research Foundation and American Water Works Association.

Solley, W.B., R.R. Pierce, and H.A. Perlman. 1993. *Estimated Use of Water in the United States in 1990.* US Geological Survey Circular 1081. Washington, D.C.: US Government Printing Office.

van der Luden, F., F.L. Troise, and D.K. Todd. 1990. *The Water Encyclopedia,* 2nd ed. Chelsea, Mich.: Lewis Publishers.

CHAPTER 6

Water Quality

It is important for all public water systems to serve their customers water that meets all established standards and is of the best possible quality. If the water quality is poor, customers are more likely to reduce consumption, and a few may turn to other sources. If customers do obtain drinking water from another source, health officials have reason to fear that the substitute could be unsafe, especially if it comes from an unprotected or contaminated source.

It is undesirable for the customer to lose confidence in the water utility and the water industry. The utility may then find it difficult to obtain approval for rate increases and hard to rally public support for improvements such as developing new water sources, upgrading treatment, and improving the distribution system.

Public loss of confidence in the utility is difficult to overcome. Confidence can be restored, however, if the utility, the regulator, environmental groups, and the media work together to reassure the public of the safety of the product. Water utility operators must take the lead in this cooperative effort by ensuring that the goals of potability and palatability are met. The actions of the utility must be open to the public, and the customers should be kept informed through bill stuffers and public service announcements. The utility may also sponsor special information programs in conjunction with publicizing National Drinking Water Week.

WATER QUALITY CHARACTERISTICS

Water quality characteristics fall into four broad categories:

1. Physical
2. Chemical
3. Biological
4. Radiological

The quality of water is based, to a great extent, on the concentration of the constituents that are present. Anything other than pure water (H_2O) is considered a constituent. In the natural environment, there is no such thing as pure water. Rainfall itself contains measurable concentrations of chemicals and particles collected from the air. As water flows above and below ground, it further acquires dissolved and suspended constituents that may make it less fit for potable use. Not all water constituents are undesirable; some are harmless, and some are even beneficial. Examples of beneficial constituents include fluoride and calcium in acceptable concentrations. Treatment is required if the levels of undesirable constituents, referred to as either impurities or contaminants, are above maximum allowable or desirable levels.

Beneficial Constituents

Not all water constituents are undesirable; some are harmless, and some are even beneficial. Examples of beneficial constituents include fluoride and calcium in acceptable concentrations.

Measuring constituent levels

The concentration of a particular constituent in water is usually very small and is measured in milligrams per liter (mg/L). This means that the quantity of a constituent in a standard volume (a liter) of water is measured by its mass (in milligrams). A concentration of 1 mg/L of magnesium in water would be about the same as 0.00013 ounces of magnesium in each gallon of water. As detection equipment has become capable of measuring concentrations at very low levels, more constituents, especially organic contaminants, are being measured and regulated at parts-per-billion—some even at parts-per-trillion—levels.

In water, concentrations expressed as milligrams per liter, within the range of 0–2,000 mg/L (i.e., dilute solution), are roughly equivalent to concentrations expressed as parts per million (ppm). For example, 12 mg/L of calcium in water is roughly the same as 12 ppm calcium in water.

Mineral concentrations

The mineral content of any water may consist of a variety of individual chemicals, such as calcium, magnesium, sodium, iron, manganese, bicarbonate, carbonate, sulfate, and chloride. In the absence of measurements of individual chemical concentrations, the mineral content of the water is determined by a laboratory process that gives the amount of total dissolved solids (not the individual constituents). The examples given in Table 6-1 demonstrate the wide range of mineral concentrations found in water bodies across the planet.

In general, river water tends to have a dissolved-minerals concentration of less than 500 mg/L. Mineral concentrations in groundwater can vary from several hundred mg/L to more than 10,000 mg/L.

Physical and Other Characteristics

The physical properties of water that are important in water treatment include:

- Turbidity
- Alkalinity
- Color
- Temperature
- Hardness
- Tastes and odors
- Dissolved solids
- Electrical conductivity

TABLE 6-1 Typical surface water mineral concentrations

Source of Water	Total Dissolved Minerals, *mg/L*
Distilled	<1
Rain	10
Lake Tahoe	70
Suwannee River	150
Lake Michigan	170
Missouri River	360
Pecos River	2,600
Ocean	35,000
Brine well	125,000
Dead Sea	250,000

Turbidity

Turbidity, or cloudiness in water, is caused by the suspended matter consisting of very fine particles that are larger than molecules. Particles that settle out very slowly, if at all, are called colloidal. Suspended material that is of inorganic origin can include clay, silt, and asbestos fibers. Suspended material that is of organic origin can include algae, bacteria, phytoplankton, and other microorganisms that are living or dead.

When a beam of light passes through turbid water, some of the light reflects off the suspended particles. Measurement of the intensity of the reflected light indicates the amount of turbidity that is present. The level of turbidity measured this way is expressed in terms of nephelometric turbidity units (ntu).

The presence of turbidity in drinking water is not only an aesthetic concern, but also a health concern. The particulates themselves may be toxic or may have contaminants adsorbed onto them. There is a particular concern that harmful microorganisms attached to the particles will be shielded against disinfectants and pose a health threat to those who drink the water.

Because of the natural filtering effect of soils, groundwater turbidity is often less than 1 ntu. However, the presence of iron and manganese may increase turbidity levels on extraction of the water. Depending on precisely when and where turbidities are measured in surface water, they can vary widely, from less than 1 ntu to more than 1,000 ntu. Because storm events wash in and stir up sediment, they can increase turbidity dramatically.

Alkalinity

Alkalinity is the measure of water's ability to neutralize acids. Alkalinity is a function of the carbonate, bicarbonate, and hydroxide content of the water. Water with higher alkalinity has higher buffering capacity—that is, the water can maintain a steady pH on addition of acid. At low alkalinity levels, even a small quantity of acid can decrease the pH dramatically. Alkalinity in source waters is influenced by minerals and salts, as well as by plant uptake and metabolism and the composition of industrial wastewater discharges. Alkalinity of water is measured by adding acid to a sample and is expressed as milligrams per liter as calcium carbonate (mg/L as $CaCO_3$). Hard waters generally have high alkalinity.

Alkalinity plays a very important role in the treatment of water. Some chemicals used to coagulate suspended matter (that is, to make it drop out of the water) consume alkalinity and therefore affect the water's pH. Sometimes chemicals are used to increase alkalinity so that sufficient pH is maintained in the treated water. Alkalinity, along with other water quality characteristics, also helps to control corrosion in water distribution systems.

Color

Color in raw water is primarily a problem for surface waters. It often indicates the presence of decomposed organic material such as leaves, roots, or plant remains. It may also be caused by a high concentration of inorganic chemicals or by upstream domestic or industrial wastewater discharges. Color in water can sometimes indicate the presence of contaminants.

Generally, color in water is reported either as true color or as apparent color, depending on the absence or presence of suspended matter, respectively, in the water sample. Color is measured by comparing the color of a water sample with the color of a standard chemical (chloroplatinate) solution. The units of measure are color units (cu). A color of less than 15 cu usually passes unnoticed, whereas a color of 100 cu has the appearance of light tea. Highly colored water is objectionable for most industrial uses; from an aesthetic standpoint, it is also unsuitable for drinking water. Figure 6-1 shows a colorimeter used to measure color.

Temperature

Water temperature, because it affects the rates of chemical reactions, is important in water treatment. Chemicals used for treatment are dissolved more easily in warm water, and suspended particles settle out more quickly. In raw surface water, warmer temperatures encourage the growth of a variety of organisms that can affect water quality.

The temperature of water in lakes generally varies with depth. During the summer, heavier cold water is normally at the bottom and warmer water is near the surface. However, as air temperature becomes colder in the fall, the surface water cools and eventually becomes slightly heavier than the water below. At that point, the lake "turns over" as the cold surface water sinks to the bottom. The turbulence created by a lake turnover can disperse decomposed bottom deposits throughout the lake. This can create a very unpleasant odor in the water and can cause quality and treatment problems for several days or weeks.

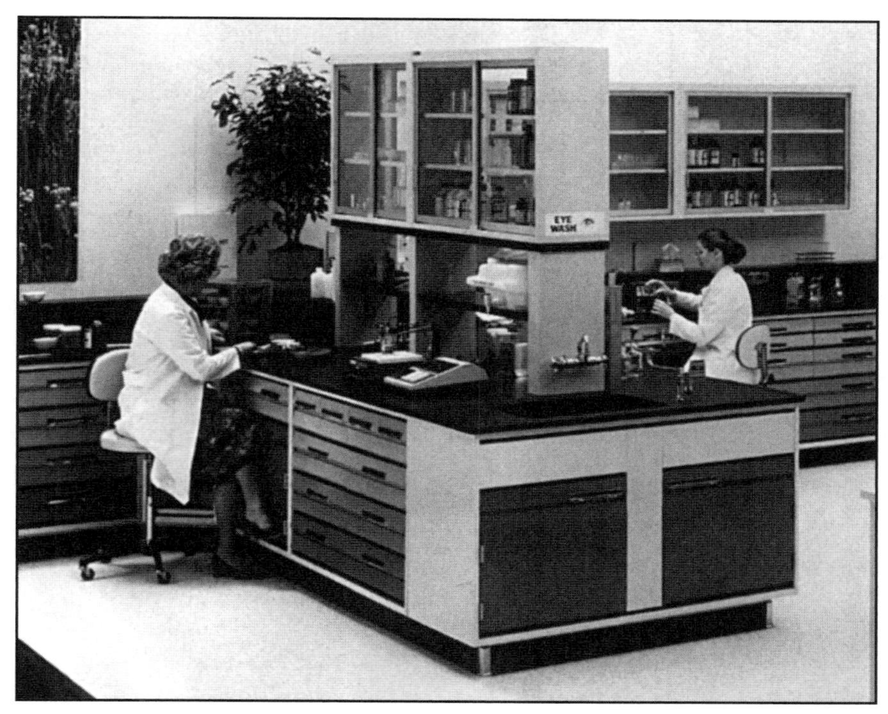

FIGURE 6-1 Colorimeter used in treatment plant lab
Courtesy of Fisher Scientific

Turnover may also occur in the spring, when ice at the surface melts into cold water that then drops down toward the lake bottom.

By knowing the water temperature at various levels in a reservoir, the operator is more likely to anticipate turnover and select the intake elevation that will deliver the best-quality water. Methods of preventing lake stratification are discussed in *Water Treatment,* another book in this series.

Water temperature is typically measured on one of two scales: Fahrenheit or Celsius. The Fahrenheit scale is primarily used in the United States. The Celsius scale is used by most other countries and generally for scientific research. Relationships between the two scales are as follows:

Fahrenheit (F): Freezing point = 32°F
Boiling point = 212°F
$$°F = (9/5)\ °C + 32$$

Celsius (C): Freezing point = 0°C
Boiling point = 100°C
$$°C = 5/9\ (°F - 32)$$

Hardness

Water is called hard when it contains significant amounts of calcium, magnesium, and other minerals. Hard water can be a problem in water supply and treatment because, like water with a high pH, it can cause scale (flaky deposits) to form in pipes and meters.

If water is hot, as in boilers and hot-water lines, the scale forms much faster. One millimeter (0.04 in.) of deposited scale can increase the cost of heating hot water by more than 10 percent. If water from a public water system is very hard, many customers install individual home water softeners.

Hard water may also have an objectionable taste and requires more soap when used for washing. A standard laboratory test for hardness determines the amount of both calcium and magnesium in the water, but for convenience, the results are expressed in milligrams per liter (mg/L) as $CaCO_3$. Hardness in water can vary from values with single digits to those in the hundreds. Table 6-2 summarizes typical degrees of hardness and the corresponding concentrations.

Tastes and odors

Undesirable tastes and odors in water can be caused by a wide variety of materials including algae, decaying organic matter, industrial and domestic wastes, minerals, and dissolved gases. Highly mineralized waters have a saline, medicinal, or metallic taste. On the other hand, small amounts of desirable minerals and dissolved gases can give water a pleasant taste. Tastes and odors become more noticeable as the water temperature increases.

Although the senses of taste and smell are closely related, the sense of smell is far more discriminating. Consequently only odor is evaluated for water treatment purposes. A specially trained person smells a series of water samples that contain increasing concentrations of the water being tested. The dilution of that sample containing the lowest detectable odor is called the threshold odor number (TON).

The TON test is conducted on raw water to determine the degree of treatment required. The test is also regularly conducted on finished water to ensure that adequate treatment is being provided. The TON in the distributed water should be less than 3. Different treatment

TABLE 6-2 Hardness concentrations and typical corresponding designations

Common Designation	Hardness, *mg/L as* $CaCO_3$
Soft	0–60
Moderately hard	61–120
Hard	121–180
Very hard	Greater than 180

strategies are adopted depending on the source of taste-and-odor-causing compounds. For example, earthy and musty odor typically caused by the presence of algae in the source water can be reduced by the use of activated carbon.

Dissolved solids

Because of its remarkable dissolving properties, water dissolves minerals from the soil and rock materials that it contacts. Table 6-3 demonstrates the wide variation in water quality that occurs in water from four different sources.

TABLE 6-3 Chemical quality comparison of four water sources

	Concentrations of Water Contaminants, *mg/L*			
Chemical	River*	Well†	Canal‡	Saline Lake§
Silica (SiO_2)	5.4	41	6.6	11
Iron (Fe)	0.11	0.04	0.11	0.10
Calcium (Ca)	9.6	50	83	2.9
Magnesium (Mg)	2.4	4.8	6.7	9.5
Sodium (Na)	4.2	10	12	8,690
Potassium (K)	1.1	5.1	1.2	138
Carbonate (CO_3)	0	0	0	3,010
Bicarbonate (HCO_3)	26	172	263	3,600
Sulfate (SO_4)	12	8.0	5.4	10,500
Chloride (Cl)	5.0	5.0	20	668
Fluoride (F)	0.1	0.4	0.2	—
Nitrate (NO_3)	3.2	20	1.3	5.8
Total dissolved solids	64	250	310	25,000

*Stream in Connecticut.
†Logan County, Colo.
‡Drainage from the Everglades in Florida.
§North-central North Dakota.

Water can at times contain many toxic dissolved minerals, including arsenic, barium, cadmium, chromium, lead, mercury, selenium, and silver. Other dissolved minerals, such as iron and manganese, can cause aesthetic problems—for example, turning water brown or black when they are oxidized.

The total quantity of dissolved solids in water is a general indicator of its acceptability for drinking and agricultural and industrial uses. Water with a high concentration of dissolved solids can create tastes, odors, hardness, corrosion, and scaling problems. In addition, high concentrations of dissolved solids have a laxative effect on most people. For these reasons, a secondary standard with a total dissolved solids (TDS) limit of 500 mg/L is recommended for drinking water.

Dissolved solids are measured by filtering a known volume of sample to remove particulates and then evaporating the filtered water to dryness. The residue that remains is weighed, and the results are recorded in milligrams per liter as filterable residue, commonly known as TDS. This test is called the filterable residue test.

Industrial users usually require water with a relatively low concentration of dissolved solids to prevent corrosion and boiler scale. If the water is to be incorporated into a product, the TDS concentration must be quite low. On the other hand, certain irrigated crops can tolerate mineralized water at a concentration as high as 2,000 mg/L.

Electrical conductivity

A common way to obtain a quick estimate of the concentration of TDS in water is to measure the electrical conductivity (EC) of the water. The greater the electrical conductivity, the greater the TDS concentration.

The instrument used to determine EC measures the electrical resistance of the water between two submerged electrodes spaced a known distance apart. The conductivity instrument readout is in micromhos per centimeter (μmhos/cm) at 25°C. As a general rule, every 10 units of EC represents 6 to 7 mg/L of dissolved solids. Therefore, an EC of 1,000 suggests a dissolved solids concentration of about 600–700 mg/L. The relationship between EC and TDS values varies depending on the precise chemical composition of the water sample.

An EC test takes only a few minutes, whereas the filterable residue test may take two hours to complete. The test is temperature sensitive and, to conform to published standards, must always be conducted with the sample at 25°C. Electrical conductivity is measured by a conductivity meter like the ones shown in Figure 6-2.

pH

The term *pH* is used to express the acidic or alkaline condition of a solution. The pH scale runs from 0 to 14, with 7 being neutral. A pH less than 7 indicates an acidic water, and a pH greater than 7 indicates a basic, or alkaline, water. The normal pH range of surface water is 6.5 to 8.5; the pH of groundwater generally ranges from 6.0 to 8.5. A typical pH meter used to determine the pH of water is shown in operation in Figure 6-3.

FIGURE 6-2 Typical electrical conductivity meters
Photo provided by YSI Incorporated

FIGURE 6-3 pH testing in the field
Courtesy of Fisher Scientific

The pH of water can have a marked effect on treatment plant equipment and processes. At a pH value less than 7, water has a greater tendency to corrode the equipment and other materials that it contacts (Figure 6-4). At a pH value greater than 7, water has a tendency to deposit scale, which is particularly noticeable in pipelines (Figure 6-5) and hot-water home appliances.

The pH of water also has an important effect on treatment processes such as coagulation and disinfection because it also affects the rates of chemical reactions. Several water treatment chemicals (e.g., corrosion inhibitors) work within a specified pH range. For finished water, to ensure corrosion control, it is desirable to have the pH somewhat higher than neutral, and the pH usually needs to be adjusted if raw- and treated-water pH values are lower than the desired value.

Dissolved oxygen

Dissolved oxygen (DO) is vital for fish and other aquatic life. Oxygen can become dissolved in water in three ways:

1. The oxygen enters the water directly from the air (through natural aeration).
2. Oxygen is introduced into the water by plant life, such as algae (through photosynthesis).
3. Oxygen is introduced by mechanical equipment during treatment, such as mechanical aeration or diffused aeration.

Dissolved oxygen is almost always present to some extent in natural waters. A warm-water lake usually contains about 5 mg/L of DO; a cold-water lake may contain 7 mg/L or more. However, if algae are present in the water, the DO level varies greatly. During sunlight hours, algae produce oxygen, so the DO level rises. During the night, algae use the oxygen, causing the DO level to drop.

Although DO is a healthy and necessary characteristic of water, it is also a significant cause of corrosion. Water containing high levels of DO corrodes the metallic surfaces of pipes, meters, pumps, and boilers. In general, the higher the DO concentration, the more rapid the corrosion will be. Dissolved oxygen can be measured either chemically or electrically. The measured concentration of DO is reported in milligrams per liter. A typical DO meter is shown in Figure 6-6.

Inorganic Constituents

Many inorganic constituents occur naturally in water, while others are found in the water as a consequence of industrial activity. Some chemicals (such as disinfectants) may enter the water as it is treated and distributed. Some inorganic constituents are associated with adverse health effects and include aluminum, antimony, arsenic, asbestos, barium, bromate, cadmium, chloride, chlorite and chlorate, chromium, copper, cyanide, fluoride, iron, lead, manganese, mercury, nickel, nitrite and nitrate, selenium, sodium, sulfates, and zinc. The actual health impact depends in large part on the concentration of the chemical in the

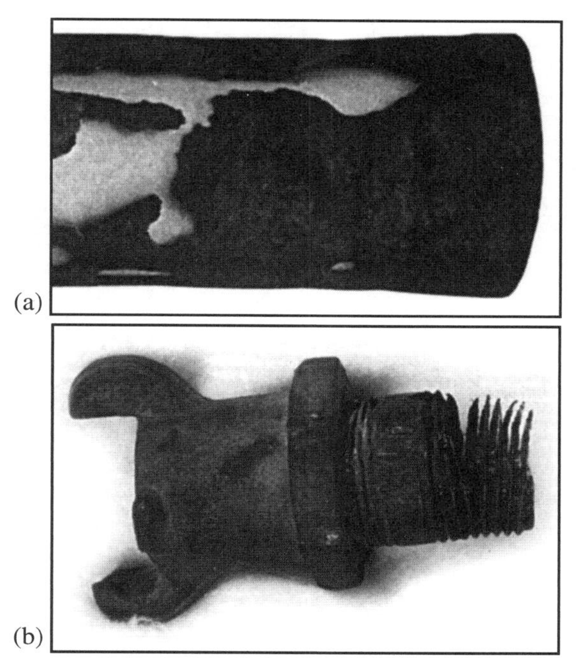

FIGURE 6-4 Corroded pipe and appurtenance

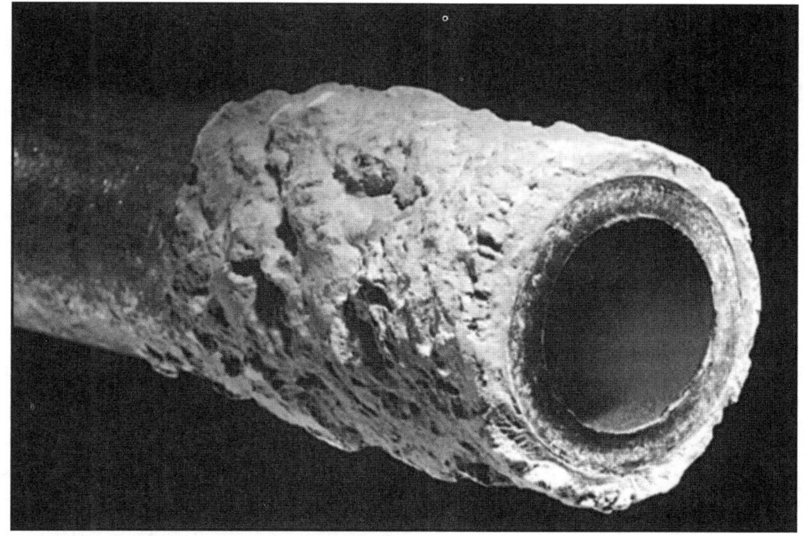

FIGURE 6-5 Pipe scaling caused by high pH values
Courtesy of Johnson Controls, Inc.

FIGURE 6-6 Dissolved oxygen meter in use
Courtesy of Fisher Scientific

water and the duration of exposure. Different methods for analyses of inorganic constituents include colorimetric, atomic absorption, inductively coupled plasma, and ion chromatography.

Organic Constituents

The concentration of organics comprise another category of chemical characteristics. To date, more than 1,000 organic chemicals have been identified in drinking water supplies. Organic chemicals in water come from several sources:

1. Plant decomposition yields materials such as tannins, lignins, and fulvic and humic materials. Surrogate parameters, such as total organic carbon (TOC) and ultraviolet (UV) absorbance, are used to quantify the natural organic matter content of water.
2. Municipal and industrial wastewater discharges may add natural and synthetic organics and pharmaceutically active compounds (PhACs).
3. Agricultural runoff yields synthetic organics such as pesticides and herbicides, as well as PhACs.
4. Water treatment operations can produce complex organics, such as trihalomethanes, during the disinfection process.

5. Runoff from roadways and parking lots carries with it oils and any chemicals that have dripped or spilled onto the pavement.
6. Medicines contribute a large part of organic chemicals. Many natural organic chemicals found in water are relatively harmless, but some synthetic organic chemicals (SOCs) can cause adverse health effects at relatively low concentrations. Some organic materials that are harmless in their natural state can react during disinfection processes to produce by-products with adverse health effects. Organic chemicals can also produce undesirable tastes, odors, and color in water. Depending on the source water characteristics, a water system is required to remove a specified level of TOC.

Biological Characteristics

The degree to which microscopic plants and animals are found in water is an important aspect of evaluating water quality. Biological constituents include viruses, bacteria, and other forms of aquatic life, as well as animal and plant contaminants.

Algae

Algae, which are present to some extent in almost all surface water, are among several varieties of aquatic plants that can seriously affect water quality. Some algae cause taste-and-odor problems; others clog sand filters or produce slime growth on equipment, tanks, and reservoir walls. Some varieties of algae can generate toxins harmful to human health. The presence of algae in source waters can affect its chemical characteristics, such as pH, and chemical changes can greatly influence the effectiveness of treatment processes within a plant. Control of algae and other aquatic plants in lakes and reservoirs is discussed in further detail in *Water Treatment*, another book in this series.

Bacteria, viruses, and other microbiological constituents

Bacteria can clog well screens, produce discolored water in the distribution system, or cause taste-and-odor problems. Of particular concern are bacteria and viruses that can cause disease in humans.

Some disease-causing microscopic organisms are not generally found in water at levels high enough to pose a health risk. However, when many of these disease-causing (pathogenic) organisms are deposited in source water by humans or animals, they can survive for a period of time and reach levels that pose a threat. Examples of such organisms are protozoa (such as *Giardia lamblia* or *Cryptosporidium parvum*) that cause intestinal diseases in humans. A healthy person may not require any special medical attention after ingesting these organisms and may recover from the parasitic infestations, but people with weakened immune systems have died as a result of the exposure. To prevent microscopic organisms from posing a health threat, disinfectants and filter membranes are used to sanitize the water prior to its distribution for consumption.

Radioactive Constituents

Radionuclides can occur in water supplies either from natural sources or as a result of human activities. Naturally occurring radionuclides include radium 226, radium 228, and radon. None of these is usually present in significant amounts in surface waters, but they do occur frequently in groundwater. Radon is a gas most often present in granite formations, where it flows through the fractures in the rock to an aquifer.

Radioactivity produced by human activities can also enter surface water supplies from a variety of sources. Potential sources include waste from medical, scientific, and industrial users of radionuclides; wastewater discharge from nuclear power plants; ash from coal-fired power plants; and discharges from the mining of radioactive materials.

FACTORS INFLUENCING SOURCE WATER QUALITY

Land use practices can have a great effect on both surface water and groundwater, and the management of those practices is an important part of source water protection.

Surface Water Quality

Surface runoff

Runoff from undeveloped areas of a watershed will not contain contaminants caused by human activities, but runoff from agricultural and urbanized land may contain chemical and microbiological contaminants that are characteristic of those land uses. Agricultural runoff can contribute animal wastes, herbicides, pesticides, and fertilizer. Urban areas contribute a wide variety of organics and inorganics from homes, streets, pet wastes, stormwater overflows, sewage outfalls, and other sources.

Point and nonpoint sources

All waste contributions to a surface water source can be classified as either point or nonpoint sources.

Simply stated, a source of water pollution is a point source if the pollutants come out of the end of a pipe connected to a particular industrial process (including wastewater treatment). The quality of wastewater from point sources is regulated by state and federal authorities under the Clean Water Act (CWA).

A source of water pollution is a nonpoint source if the pollutants enter the water is a result of runoff from the lands surrounding the water body without going through a well-defined entry point. These sources of water contamination are much harder to control because the generation of pollution occurs across many properties over a wide area and in diffused form.

Water can become contaminated in many ways, and contamination at one location may be caused by contamination sources many miles away. The factors that influence source water quality may be categorized as natural or human.

Natural factors include:

- Climate
- Watershed characteristics
- Geology
- Microbial growth
- Fire
- Saltwater intrusion
- Density (thermal stratification)

Human factors associated with point sources include:

- Wastewater discharges
- Industrial discharges
- Hazardous-waste facilities
- Mine drainage
- Spills and releases

Human factors associated with nonpoint sources include:

- Agricultural runoff
- Livestock
- Urban runoff
- Land development
- Landfills
- Erosion
- Atmospheric deposition
- Recreational activities

Water from reservoirs

When runoff enters a lake or storage reservoir, its quality changes in several ways, some of them beneficial and others not beneficial. As water enters a reservoir and slows down substantially, suspended material in the flowing water settles. Bacterial action and natural oxidation can act on soluble contaminants and suspended particles to make them less likely to cause color and taste-and-odor problems; sometimes the soluble material may be converted into a particulate form that will settle.

If a reservoir is large, it can absorb and dilute the surges of particularly turbid runoff carried by streams following a storm event. It may even be able to absorb and dilute a quantity of poor-quality water or a small chemical spill brought in by a stream. Wind action and algae help keep the dissolved oxygen in a reservoir at beneficial levels, and that, in turn, both promotes natural purification processes and supports aquatic life.

Most bacteria that are pathogens, as well as protozoa that cause disease, do not naturally occur in fresh water in significant amounts, in the United States. Although some bacteria can survive for a fairly long time after leaving a human or animal, they cannot multiply effectively. When held in a reservoir for a period of time, many disease-causing organisms become nonviable and noninfectious.

There are three general categories of algae: diatoms, green algae, and blue-green algae. Because they all contain chlorophyll, they grow and multiply when exposed to light in the presence of adequate nutrients.

Diatoms and green algae are found in reservoirs that have low nutrient levels, and they help keep the water oxygenated. If they "bloom," or grow in profusion, they can clog intake screens and filters. At times, they can also cause taste-and-odor problems.

Blue-green algae are more commonly found in nutrient-rich waters where they can bloom profusely, especially when water temperatures are higher. Blue-green algae are notorious for causing taste-and-odor problems. Even though algae produce oxygen while living, the die-off and subsequent decomposition of a sizable bloom can have the opposite effect, depleting the water of oxygen and making it unfit for other forms of aquatic life.

All lakes progress through a natural process called succession. In the early stages of succession, nutrient levels and the corresponding rates of biological productivity are relatively low; little aquatic life will be found. As more nutrients enter the lake, it becomes more productive (or increasingly more eutrophic). Organic material and sediment accumulate, and eventually the lake becomes swampy or marshy. This succession process can take many decades, even centuries, depending on size of the water body. The artificial introduction of nutrients into a lake from human activities can speed up the process. Heavy loads of nutrients cause eutrophication of the lake, a process that causes it to be overproductive. Eutrophication typically results in more frequent and more substantial algae blooms, and it may cause a lake to become unusable as a water source. Efforts to prevent or reverse eutrophication may include stopping the inflow, applying a chemical treatment so nutrients settle out, or aerating the water.

Additional details on treatment of lakes and reservoirs for algae and water weed control can be found in *Water Treatment,* another book in this series.

Atmospheric Contaminants

Air pollution can have a significant effect on water quality because chemicals and particulates in the air eventually fall out with precipitation. One notable effect of air pollution on water resources is acid deposition (both wet and dry depositions).

The most common component of acid deposition is wet deposition (i.e., acid rain) that contains sulfuric and nitric acids. These acids are produced in the atmosphere from the combustion of coal for power generation. Much coal naturally contains sulfur, and when the coal is burned, sulfur dioxide is among the chemical compounds that are released into the atmosphere, along with nitrogen oxides through the combustion process. A process known as photochemical oxidation then converts sulfur dioxide in the air to sulfuric acid. This acid can be carried hundreds of miles and then deposited as acid rain, acid snow, and even acid dust. The pH in many lakes in the northeastern United States and Canada has been reduced due to acid rain. The change in pH not only has a direct effect on aquatic life, but it also can change the chemistry of the water by causing mobilization and higher concentrations of dissolved minerals in the water. Where acid rain diminishes the health of natural vegetation, increased erosion can result.

Other sources of airborne contaminants are automobile exhaust, mercury from coal-fired power plants, emissions from waste incinerators, volcanic eruptions, and almost any combustion process that does not include complete treatment of stack gases. Title IV of the 1990 Amendments to the Clean Air Act (CAA) mandates emissions requirements to control acid rain and has significantly reduced the severity and extent of the problem.

Groundwater Quality

Groundwater is not as vulnerable to direct pollution and contamination as surface water, but contamination can become a serious problem, as with leaky underground storage tanks, chemical spills, failure of on-site waste treatment systems, underground injection of wastes, inadequate well protection, leaching from landfills, and overfertilization of crops.

Aquifer contamination

Because of the protective soil cover and the natural filtration provided by aquifer material, the biological characteristics of groundwater are generally very good. Because harmful bacteria do not penetrate very far into the soil, wells more than 50 ft (15 m) deep are generally free of harmful organisms. Some exceptions to this are aquifers with large voids in fractured rock or aquifers with conduits (such as improperly abandoned wells) that provide pathways for contaminants to enter the aquifers.

Many cases of groundwater contamination (Figure 6-7) are a result of either poor well construction or poor waste disposal practices on the land overlying the aquifer. Other sources of contamination may include recharge from nearby surface waters or excessive pumping near coastal areas, resulting in saltwater intrusion into the groundwater resources. The movement of water in the ground can treat water naturally. Many water supply systems, especially in European countries, employ riverbank-filtered water as their source of supply.

Chemical Characteristics

Groundwater acquires its chemical characteristics in two ways: underground from the soil and rock that it contacts and from the surface water that percolates into the aquifer. Typically, aquifers are recharged by the rain that falls on the aquifer recharge area. This water is generally free of anthropogenic contaminants if the rain falls on land that is free from pollution; but the water can become contaminated if it carries undesirable chemicals from the

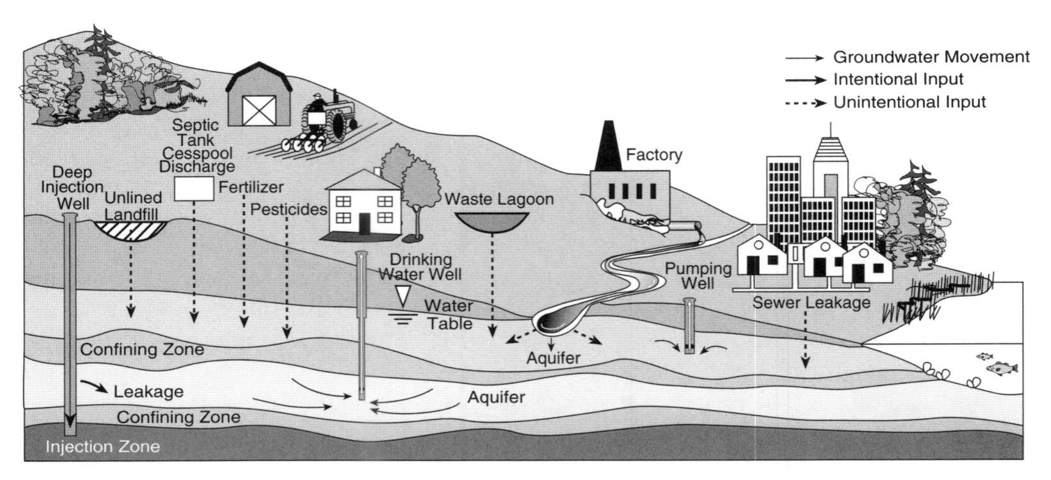

FIGURE 6-7 Sources of groundwater contamination

Source: Monitoring and Assessing Water Quality, National Water Quality Inventory: 1998 Report to Congress, Chapter 7, page 6, USEPA.

landscape. In agricultural areas, for example, water that percolates down into an aquifer can be contaminated with nitrates, herbicides, or pesticides.

Not all contamination of groundwater comes from surface contamination. In coastal areas, saltwater intrusion presents a major concern. Under natural conditions, the boundary between subsurface fresh water and subsurface salt water near the coast is generally in equilibrium and moves very little. When well operation removes the fresh water faster than it is replenished, the salt water moves underground toward the wells. Where saltwater intrusion is a threat, wells may still be used, but withdrawal rates must be limited to prevent saltwater from reaching the wells.

Several natural contaminants affect the potability of groundwater. Some elements and compounds are beneficial or harmless in low concentrations; they become troublesome or harmful only when found in high concentrations.

Some contaminants, such as barium, arsenic, and various radionuclides, are harmful at even very low levels. Gases that are undesirable in groundwater include hydrogen sulfide, methane, and radon. Common groundwater contaminants and their associated problems are listed in Table 6-4.

Chemical contaminants

Synthetic organic compounds in the vicinity of public water supply wells are a very serious concern. Chemicals that have been carelessly disposed of on the ground or buried in unlined pits have been found in many well supplies. Some of these chemicals are considered a health threat at concentrations far less than can be detected by taste or smell. The only way they can be detected is by relatively complicated and expensive tests.

TABLE 6-4 Common groundwater quality problems

Constituent	Source	Type of Problem	USEPA Primary Drinking Water Standards*	USEPA Secondary Drinking Water Standards†
Inorganic Constituents				
Arsenic	Naturally occurring‡	Toxic	0.05 mg/L	—
Fluoride	Naturally occurring	Stains teeth, can cause tooth damage at high levels	4.0 mg/L	—
Hydrogen sulfide	Naturally occurring	Offensive odor, flammable, corrosive	—	0.05 mg/L
Iron	Naturally occurring	Stains plumbing fixtures and laundry, causes tastes	—	0.3 mg/L
Manganese	Naturally occurring	Discolors laundry and plumbing fixtures, causes tastes	—	0.05 mg/L
Nitrate	Fertilizer and fecal matter	Toxic to infants	10 mg/L (as nitrogen)	—
Radioactivity	Naturally occurring	Cancer causing	Gross alpha activity (15 pCi/L) Radium 226 and 228 (5 pCi/L)	—
Sodium	Naturally occurring	May contribute to high blood pressure	—	Being investigated
Sulfate	Naturally occurring‡	Laxative effect	—	250 mg/L
Total dissolved solids	Naturally occurring‡	Associated with tastes, scale formation, corrosion, and hardness	—	500 mg/L

Table continued next page

TABLE 6-4 Common groundwater quality problems (Continued)

Constituent	Source	Type of Problem	USEPA Primary Drinking Water Standards*	USEPA Secondary Drinking Water Standards†
Organic Constituents				
Pesticides and herbicides	Agricultural and industrial contamination	Many are toxic; cause tastes and odors	Several are regulated	—
Solvents	Industrial contamination	Many are toxic; cause tastes and odors	Being developed	—
Microbiological Constituents				
Disease-causing microorganisms	Fecal contamination	Cause variety of illnesses	Coliform bacteria are regulated as indicator organism	—
Iron bacteria	Contamination from surface	Produce foul-smelling slimes, which plug well screens, pumps, and valves	—	—
Sulfate-reducing bacteria	Contamination from surface	Produce foul-smelling and corrosive hydrogen sulfide	—	—

Source: Basics of Well Construction Operator's Guide.
Note: USEPA = US Environmental Protection Agency
*State regulations may be more restrictive.
†Not enforceable at federal level.
‡The naturally occurring elements can also be present as a result of contamination by humans.

One common contaminant that makes groundwater unfit to drink is gasoline, which is a complex mixture of toxic organic compounds. Thousands of wells, mostly private individual home wells (but some public supplies, too), have been contaminated because of leaking underground gasoline tanks. Less widespread, but not uncommon, is contamination from underground fuel oil storage.

If chemical contamination of a well is detected, it is generally best to abandon the well and use another water source. If no alternative source is available, the chemicals can usually be removed from the water by aeration or carbon adsorption processes. But it is much more desirable to prevent groundwater contamination in the first place than to have to treat the contaminated water indefinitely.

PUBLIC HEALTH SIGNIFICANCE OF WATER QUALITY

Water quality can have a significant effect on public health as a result of the following:

- Waterborne diseases
- Inorganic chemical contaminants
- Disinfectants and disinfection by-products (DBPs)
- Organic chemical contaminants
- Radioactive contaminants

Waterborne Diseases

Outbreaks of waterborne diseases are usually caused by contact with contaminated source water, ineffective treatment operations, and/or contamination in the water distribution system. Even though modern treatment removes or inactivates known disease-causing organisms sufficiently to keep their occurrence at safe levels, it is better to have a source of raw water that is uncontaminated in the first place. Waterborne diseases are generally acute and of short duration, and symptoms can be similar to those of digestive maladies caused by factors such as contaminated, uncooked, or improperly cooked food. Table 6-5 lists the causes and health effects of common waterborne diseases.

The sources of disease-causing organisms in water include combined sewer overflows, sanitary sewer overflows, wastewater treatment plant upsets, agricultural operations, surface runoff containing human or animal fecal contamination, or contributions from an on-site waste treatment system (e.g., septic tanks). The single most prevalent waterborne disease is giardiasis. *G. lamblia*, the cause of the disease, is a protozoan that can infect both humans and animals. Two common carriers are beavers and muskrats, which, by their very nature, are commonly found around surface water bodies. The life cycle of *G. lamblia* includes a cyst form that can remain viable under adverse conditions in water for one to three months. The cysts are quite resistant to chlorine and can pass through filters that are not operating efficiently. Cryptosporidiosis, caused by another protozoan known as *Cryptosporidium*, is associated with most reported cases of illness due to waterborne outbreaks.

TABLE 6-5 Potential waterborne disease-causing organisms

Organism	Major Disease	Primary Source
Bacteria		
Salmonella typhi	Typhoid fever	Human feces
Salmonella paratyphi	Paratyphoid fever	Human feces
Other *Salmonella* sp.	Gastroenteritis (salmonellosis)	Human/animal feces
Shigella	Bacillary dysentery	Human feces
Vibrio cholera	Cholera	Human feces, coastal water
Pathogenic *Escherichia coli*	Gastroenteritis	Human/animal feces
Yersinia enterocolitica	Gastroenteritis	Human/animal feces
Campylobacter jejuni	Gastroenteritis	Human/animal feces
Legionella pneumophila	Legionnaires' disease, Pontiac fever	Warm water
Mycobacterium avium intracellulare	Pulmonary disease	Human/animal feces, soil, water
Pseudomonas aeruginosa	Dermatitis	Natural waters
Aeromonas hydrophila	Gastroenteritis	Natural waters
Helicobacter pylori	Peptic ulcers	Saliva, human feces?
Cyanobacteria	Gastroenteritis, liver damage, nervous system damage	Natural waters
Enteric Viruses		
Poliovirus	Poliomyelitis	Human feces
Coxsackievirus	Upper respiratory disease	Human feces
Echovirus	Upper respiratory disease	Human feces
Rotavirus	Gastroenteritis	Human feces
Norwalk virus and other caliciviruses	Gastroenteritis	Human feces

Table continued next page

TABLE 6-5 Potential waterborne disease-causing organisms (Continued)

Organism	Major Disease	Primary Source
Hepatitis A virus	Infectious hepatitis	Human feces
Hepatitis E virus	Hepatitis	Human feces
Astrovirus	Gastroenteritis	Human feces
Enteric adenoviruses	Gastroenteritis	Human feces
Protozoa and other organisms		
Giardia lamblia	giardiasis (gastroenteritis)	Human and animal feces
Cryptosporidium parvum	Cryptosporidiosis (gastroenteritis)	Human and animal feces
Entamoeba histolytica	Amoebic dysentery	Human feces
Cyclospora cayatanensis	Gastroenteritis	Human feces
Microspora	Gastroenteritis	Human feces
Acanthamoeba	Eye infection	Soil and water
Toxoplasma gondii	Flulike symptoms	Cats
Naegleria fowleri	Primary amoebic meningoencephalitis	Soil and water
Fungi	Respiratory allergies	Air, water?

Source: Water Quality and Treatment, 5th ed., *1999, AWWA.*

Cryptosporidium parvum, the species that causes disease in humans, produces oocysts that are smaller in size than the *Giardia* cyst and are also more resistant to chlorine disinfection. However, filtration and the use of other disinfectants (e.g., ozone, chlorine dioxide, and UV light) provide effective control.

Most reported outbreaks that are traceable to a water supply originate from unfiltered water systems or contaminated water sources (surface water as well as groundwater). Outbreaks may also occur when water system filters are not functioning properly.

Organisms that cause waterborne disease can be extremely difficult to detect directly. Only very well equipped laboratories may be able to perform the tests for some specific organisms. The presence of microbial contamination is detected by analyzing for surrogate microbial parameters (or indicators) such as total coliforms, heterotrophic plate counts, or

Escherichia coli. Positive detection of these parameters may indicate possible contamination and a need for further evaluation. Confirmed contamination may require public notification and the implementation of deliberate steps to identify and eliminate the causes of contamination.

Inorganic Chemical Contaminants

Among inorganic contaminants that can be found in water, metals are a major concern, but nonmetallic compounds can also pose serious problems. Some inorganic contaminants (e.g., arsenic) are associated with source water, and others (e.g., lead and asbestos) might enter the water during distribution. Some inorganic contaminants are suspected carcinogens (e.g., lead and arsenic).

The National Primary Drinking Water Regulations, enforced by the EPA, specifies maximum contaminant levels (MCLs) for a number of inorganic chemicals that may be present in drinking water. For more information, see www.epa.gov/safewater/contaminants/index.html.

Disinfectants

Other disinfectants in common use are chlorine dioxide, ozone, and potassium permanganate. These chemicals have some advantages for use in drinking water disinfection. Their principal disadvantage is that they do not provide a disinfectant residual that will remain active against any bacteria in the water for a period of time once the water enters the distribution system. Therefore, chlorine in the free or combined form continues to be the chemical of choice for maintaining disinfectant residual in the distribution system. As part of the Stage 1 Disinfectant/Disinfection By-Products (D/DBP) Rule, which became effective in January 2002 for systems serving more than 10,000 persons, disinfectant residuals will be controlled in the finished water to minimize the risk of adverse health effects.

Other methods of disinfecting water include the use of iodine, bromine, silver, and hydrogen peroxide. However, these methods are rarely used to disinfect public drinking water supplies.

Chlorine

Chlorine is an inorganic chemical that is used widely and effectively as a disinfectant. All forms of chlorination, including gaseous chlorine, liquid chlorine, sodium hypochlorite, and calcium hypochlorite, produce hypochlorous acid in dilute water solution. For killing or inactivating pathogens, hypochlorous acid is the most effective of the various chlorine compounds that are produced when water is chlorinated. Undiluted, all chlorine disinfectants are strong chemicals and are toxic to plants and animals. At the low concentrations used for water disinfection, however, they are considered to have very low toxicity.

Free chlorine reacts with some of the natural organic compounds present in raw water to form varying amounts of chemicals that fall into a group called trihalomethanes

(THMs) and haloacetic acids (HAAs). The principal THM is chloroform. These compounds, as a group, are considered carcinogenic. Consequently, state and federal regulations limit the amount of THMs that may be present in water provided to the public.

Because the disinfection by-products (DBPs) from chlorination may cause adverse health effects, chlorinated water should be monitored for unhealthy concentrations of DBPs. The risk of adverse health effects from DBPs in chlorinated water is very low, however, and the health benefits are consistently much higher. Consequently, chlorination of public water supplies, as required by regulations, is commonplace in the United States. Utilities must comply with the USEPA's D/DBP Rule.

Chloramine

Chloramine is a disinfectant that is formed when chlorine reacts with ammonia in the water. Chloramine reduces tastes and odors in some water and produces a lower concentration of THMs, but it does not have the disinfecting power of free chlorine against bacteria. It is also considered relatively ineffective for inactivating *Giardia*.

Ultraviolet light

Ultraviolet disinfection technology, which requires no chemicals, can be used to inactivate both *Giardia lamblia* and *Cryptosporidium parvum* in drinking water. Interest in using UV light has been growing among public water supply systems because it does not result in disinfection by-products. Guidance for using UV to meet regulatory standards has been published by the USEPA in its *Ultraviolet Disinfection Guidance Manual* (2006).

Ozone

Ozone, a chemical comprised simply of three oxygen atoms, is a very powerful oxidant and disinfectant. Ozone is produced by generating an electrical discharge (or spark) through an oxygen-rich gas, and it must be generated on-site. Ozone is widely used in Europe and has been growing more popular in the United States. It destroys organic matter and is very effective against bacteria, viruses, *Giardia*, and *Cryptosporidium*. It quickly decomposes in water, however, and consequently does not have a protective residual effect in the distribution system. When ozone is used as an initial disinfectant for water, chlorine must then be added to provide a disinfectant residual. Both the initial capital costs for equipment and the operating costs of using ozone are considerably higher than costs for chlorine.

Organic Chemical Contaminants

Organic contaminants in drinking water come from the following three major sources:

1. The breakdown of naturally occurring organic materials
2. Contamination from industrial, commercial, and domestic activities
3. Reactions that take place during water treatment and distribution

The first source is by far the most significant. It consists of products that result from the decay of leaves and other plant material, aquatic animal death and decomposition, algae decomposition, and other aquatic life by-products. Most of these organics, except for some highmolecular aliphatic and aromatic hydrocarbons, have no direct adverse health effects. Upon the chemical disinfection of water, however, some of these organics react to produce DBPs, which do have potential adverse health effects.

Organic chemicals can be found in surface water as a result of agricultural, industrial, commercial, and domestic activities. They can get into the water through sewage discharges and groundwater seepage, or by means of runoff from contaminated land and urban pavement. This group of chemicals includes oils, fuels, solvents, pesticides, herbicides, and other organics used by business, industry, and the public. There are regulations covering the use and disposal of many of these chemicals, and many of these organic compounds are specifically regulated in drinking water.

Polychlorinated biphenyls (PCBs), used in a variety of manufacturing processes, comprise one of the most notorious groups of organic chemicals. Laboratory tests have shown that PCBs can cause cancer, and they can also have harmful effects on the immune system, reproductive system, nervous system, and endocrine system. Discharge of PCBs in industrial wastes and evidence of their presence in the environment generated immense concern in the 1970s and ultimately led to the banning in 1979 of all PCB manufacturing in the United States.

Another chemical that has gotten a great deal of public attention is dioxin, which appears to be a potent animal carcinogen. The principles, occurrence, and control of organic chemicals in drinking water are covered in detail in *Water Quality*, part of this book series.

Several organic compounds referred to as endocrine disruptors can have damaging effects on humans and other animals by interfering with the production of hormones. The USEPA maintains an Endocrine Disruptor Screening Program to identify compounds that require regulation to keep concentrations of these chemicals below harmful levels.

Radioactive Contaminants

The radioactivity found in water is almost always from natural sources. Groundwater in particular may contain radionuclides. All radioactive contaminants are to some degree carcinogenic. The radionuclides of concern in drinking water are:

- Uranium
- Gross alpha and beta emitters
- Radium 226
- Radium 228
- Radioactive cesium, iodine, and strontium 89 and 90
- Radon
- Thorium 230
- Thorium 232

The SI unit commonly used to express and compare the biological effects of radiation from different sources is Sievert (Sv). (The older unit is the rem; 1 Sv is equal to 100 rem.) At low doses, the milliSievert (mSv) is used.

Human beings are constantly exposed to a small amount of background radiation from cosmic rays and the soil and rocks in the environment. The average annual human dose from background sources is about 2.4 mSv per year, but the dose can vary considerably by location. The US Environmental Protection Agency (USEPA) estimates that radioactivity in drinking water contributes about 0.1 to 3 percent of the annual dose received by an average person. The radionuclide that is usually responsible is radon, and radon levels can vary drastically from one location to the next.

It is estimated that between 500 and 4,000 public water supplies have radon at levels that create an equivalent dose of 1 mSv per year. In addition, about 30,000 water systems have radon at a level at which some human risk can be calculated.

PUBLIC WATER SUPPLY REGULATIONS

At any one utility, drinking water is regulated by a combination of federal, state, and local agencies. Regulations for operating and monitoring water systems serving the public were initially developed by states in the early 1900s, but the policies and the degree of enforcement varied considerably among the states. Before 1974, federal involvement in regulating the provision of drinking water to the public consisted only of the development of drinking water standards by the US Public Health Service. These standards suggested upper limits for the presence of bacteria and chemicals in water from public water systems. The first federal drinking water standard was passed in 1914 to control *E. coli* in water supplies used by interstate carriers. In 1925, additional standards were developed for physical and chemical constituents in drinking water. By 1962, 28 contaminants were regulated, but these standards were applicable to less than 2 percent of US water supply systems. Although the standards were not federally enforceable, they were incorporated into the regulations of most states. Each state's health department generally monitored water quality, enforced regulations, and provided technical assistance to water system operators.

The Safe Drinking Water Act

During the 1960s and early 1970s, scientists and public health experts determined more fully the potential for harmful disease-causing organisms and chemicals in drinking water. There was also increasing pressure by the public and legislators to create uniform national standards for drinking water to ensure that every public water supply in the country met minimum health standards.

In response to these concerns, Congress passed the Safe Drinking Water Act (SDWA) in 1974. The SDWA directs the USEPA to establish standards and requirements necessary to protect the public from all known harmful contaminants in drinking water. It also asks each state to accept primary enforcement responsibility (primacy) for enforcing the federal requirements. Since its inception, the SDWA has been amended, or reauthorized, several times; the most significant amendments were made in 1986 and 1996. The original SDWA and the 1986 amendments focused on treatment in order to provide safe drinking water. The 1996 amendments included source water protection, operator training, and funding for water system improvements as important aspects of providing safe drinking water.

SDWA provisions

Many requirements in the SDWA relate to public water system operation, monitoring, and reporting. The more important points of the regulations are as follows:

- Based on available sound science and cost-effective technologies, the USEPA establishes primary drinking water standards for microbiological and chemical contaminants that may be found in drinking water and that could cause adverse health effects for humans. The maximum concentration that is allowable is called the maximum contaminant level. All states must make their regulations at least as stringent as the MCLs established by the USEPA. The primary standards are mandatory, and all public water systems in the United States must comply with them.
- The USEPA establishes secondary drinking water standards that set recommended standards for contaminants that are not a direct threat to public health but affect aesthetic qualities such as tastes, odors, color, and staining of fixtures. The secondary standards are strongly recommended but not enforceable.
- SDWA mandates that states have certification programs and recertification requirements for water system operators and that they ensure small water systems have the technical, financial, and managerial capacity to provide safe drinking water.
- Public notification is required by the SDWA as the first step of enforcement against all public water systems that fail to comply with federal requirements. Systems that violate operating, monitoring, or reporting requirements or that exceed an MCL must inform the public of the problem and explain the public health significance of their violation. In addition, SDWA requires all community water systems to prepare and distribute annual consumer confidence reports, including sources of water and detected contaminants and their associated health effects.
- SDWA mandates that each state conducts an assessment of its sources of drinking water to identify potential sources of contamination.
- Federal enforcement, including stiff monetary fines, may be leveled against water systems that do not comply with the federal requirements.
- A public water system is defined as a system that supplies piped water for human consumption and has at least 15 service connections or 25 or more persons who are served by the system for 60 or more days a year.

Provisions of the SDWA and related regulations enacted by the USEPA are explained in detail in *Water Quality*, which is also part of this series.

Classes of public water systems

To implement the SDWA, the USEPA has classified public water systems into three categories based on the number and type of customers served:

1. **Community public water systems** are water utilities that serve the same customers year-round. Examples are municipal systems, rural water districts, and mobile home parks.
2. **Nontransient, noncommunity public water systems** are entities that have their own water supply that serves the same customers for more than six months out of the year. Examples include schools, factories, and office buildings.
3. **Transient, noncommunity public water systems** have their own water system that serves customers but not the same individuals for more than six months a year. Examples include parks, motels, campgrounds, restaurants, and churches.

Community and nontransient, noncommunity water systems are required to meet specific operating requirements and to perform relatively thorough monitoring of their water quality. This is necessary because their customers use the water over a long period of time and could experience adverse health effects from continuous exposure to relatively low levels of contaminants. The requirements for transient, noncommunity systems are not as stringent because the customers use the water only occasionally.

Water systems having fewer than 25 customers and individual homes having their own source of water (such as a well) are referred to as private or nonpublic water systems. Nonpublic systems are not covered by the SDWA but are usually regulated by state and/or local jurisdictions.

State Drinking Water Programs

The intent of the SDWA is for each state to accept primary enforcement responsibility for operation of the drinking water program within the state. To be given primacy, a state must implement all requirements, standards, and programs that are required by the USEPA. In return, the state receives a federal grant to supplement state funds to operate the public water supply program.

Most states also have their own programs. These include certification of water system operators, cross-connection control, and monitoring for contaminants that are not part of the federal requirements. Utility operators will find official guidance for these programs within their respective state agencies.

The major functions provided by state drinking water agencies that exercise primacy under the SDWA are as follows:

- **Monitoring and tracking.** The state staff must ensure that all systems perform proper monitoring, use approved laboratories, and regularly submit reports on system operation.
- **Sanitary surveys.** A sanitary survey is an on-site inspection of a water system's facilities and operation. Surveys are performed by state staff or other qualified persons on

a regular basis. Reports are prepared for the system owner or operator and problems or deficiencies are identified.

- **Plan review.** The SDWA requires that states review and approve all plans for water system construction and major improvements.
- **Technical assistance.** The state provides water system owners and operators with technical assistance when they have problems.
- **Laboratory services.** Chemical, radiological, and microbiological analyses of water quality must be performed by a certified laboratory. In some states, state laboratories perform all or most of the tests. Other states certify commercial laboratories that do the work for the water systems.
- **Enforcement**. The SDWA requires states to use enforcement actions when federal requirements are violated.

As the federal requirements are periodically changed and expanded, each state must, within a prescribed period of time, make similar changes in its regulations and the corresponding implementation and enforcement policies.

SELECTED SUPPLEMENTARY READINGS

Basics of Well Construction Operator's Guide. 1986. Denver, Colo.: American Water Works Association.

Manual M7, *Problem Organisms in Water: Identification and Treatment.* 2003. Denver, Colo.: American Water Works Association.

Manual 48, *Waterborne Pathogens.* 2006. Denver, Colo.: American Water Works Association.

Manual 57, *Algae.* 2010. Denver, Colo.: American Water Works Association.

Manual of Instruction for Water Treatment Plant Operators. 1975. Albany: New York State Department of Health.

Monitoring and Assessing Water Quality, National Water Quality Inventory: 1998 Report to Congress. USEPA. Available: www.epa.gov/305b/98Report/Chap7.pdf.

Scharfenaker, M., J. Stubbart, and W.C. Lauer. 2006. *Field Guide to SDWA Regulations.* Denver, Colo.: American Water Works Association.

VOCs and Unregulated Contaminants. 1990. Denver, Colo.: American Water Works Association.

Water Quality and Treatment, 6th ed. 2010. New York: McGraw-Hill and American Water Works Association (available from AWWA).

CHAPTER 7

Water Source Protection

The quality of source water is subject to threats both from nature and from human activities. Natural threats include saltwater intrusion, errant microbiological growth, and fire. Human threats can be categorized as point or nonpoint sources. Point sources include wastewater treatment plant discharges, hazardous-waste facilities, mine drainage, and spills and releases. Nonpoint sources include agricultural and urban runoff, landfills, erosion, and recreational activities (such as boating on reservoirs). It is the utility's responsibility to minimize harm from both natural and human threats. Both surface water and groundwater may be contaminated by human activity.

The US Environmental Protection Agency (USEPA) web site (www.epa.gov) contains valuable information regarding protection of sources of both raw groundwater and surface water. Likewise, state governments should have a substantial amount of useful information about source protection, because the Safe Drinking Water Act (SDWA) amendments of 1996 require each state to develop a Source Water Assessment Program.

FUNDAMENTAL PRINCIPLES

Surface water can usually recover from pollution incidents much more quickly than groundwater. However, surface water is also more vulnerable to adverse natural and human influences. Some threats to water sources are relatively obvious—for example, the installation of a gasoline storage tank near a well or shoreline grazing of a herd of cattle along a small reservoir. Unfortunately, most threats are not this obvious, and many are difficult to identify, even with time and hard work.

SURFACE WATER PROTECTION

The mandate that a public water utility must provide wholesome water means that its interests begin not at the raw water intake, but extend upstream to the entire watershed. As an example of the comprehensive nature of surface water protection, the Massachusetts Watershed Protection Plan provides the following statement, which applies to all watersheds, regardless of whether or not the water utility will be filtering the water:

> The protection of watersheds for ensuring the public's right to clean drinking water will require some creative and innovative approaches involving the cooperation of local, regional, and state levels of government. No agency alone can achieve the ambitious objectives required, because watersheds cover expansive areas, often encompassing several communities, and often include bordering states. Therefore, it will be in the public interest that all interested parties participate at each level of government and develop the cooperative mechanisms locally, regionally, and at the state level. The issue will be our greatest challenge and our greatest threat.

Source Protection Area

The first step in source protection is to determine the geographic area that needs protection and determine who exactly has an interest in protecting it. All of the water system's customers have an obvious stake in their water supply and should be kept informed of potential threats to water quantity or quality. Landowners in the protection area may be legally or financially affected by contamination prevention requirements, so they should certainly be involved. Municipal and county officials may have to act as an enforcement arm, so they should be consulted, and public health and environmental protection agencies should also be kept informed. When all interested parties have been contacted and, optimally, when a task force has been formed, the effort to protect the water supply in a fair and responsible way can commence.

Watershed mapping

If a water utility uses a surface water source, the watershed boundary (or drainage area) above the point of withdrawal should be delineated. The outside boundary of a watershed is the surface water divide between watersheds. As illustrated in Figure 7-1, various types of maps can be used in watershed mapping and for identifying surface features. The most accurately drawn watersheds use topographical maps like the one shown in Figure 7-2. US Geological Survey (USGS) topographic maps are particularly useful for outlining a watershed. The ridges between watersheds are easily identified. Water on one side of a ridge flows toward one watershed, and water on the other side flows in another direction. Using lines to connect the ridges on the map will outline the area contributing water to the source used by a water system.

Watershed delineation can also be done quickly and reliably using desktop computers and mapping software. The software can read digital files that contain elevation data and automatically determine the drainage area above any given point on the landscape.

Once a watershed has been identified, the topographic maps can also be used to identify the general types of vegetation and activities that exist within the drainage area. Features such as forests, open land, roads, industrial activities, and houses that were present when the map was prepared are all shown. These data can be useful in determining the streamflow characteristics in addition to potential sources of contamination in the watershed.

Digital geographic information system (GIS) data about land use, land cover, and local zoning can also provide an overview of activities in the drainage area above a particular intake. Digital aerial photographs can be lined up with other data to provide an accurate and current picture of the watershed.

Watershed Resource Protection Plan

A watershed resource protection plan generally consists of the following elements:

- A delineation of the watershed
- A description of the watershed's topography, soils, and land cover

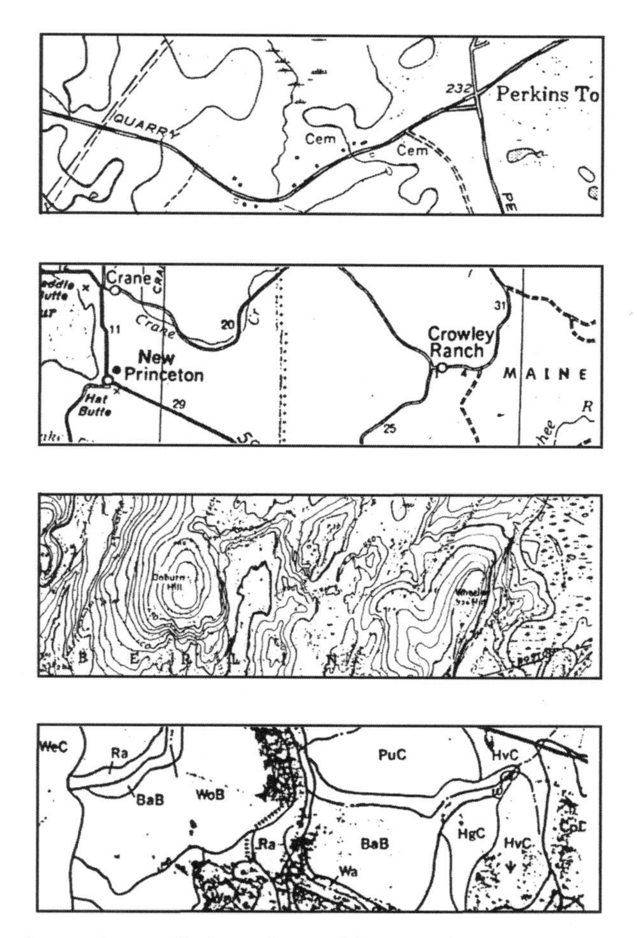

FIGURE 7-1 Various types of maps that can be used in watershed mapping
Source: Protecting Local Ground-Water Supplies Through Wellhead Protection (1991).

- Identification of land uses, highlighting those that could be detrimental to water quality
- An assessment of the risks and how best to control detrimental activities and events
- How detrimental activities and events should be monitored
- Determination of land ownership—what agreements are needed to control detrimental activities
- How best to manage and operate the program

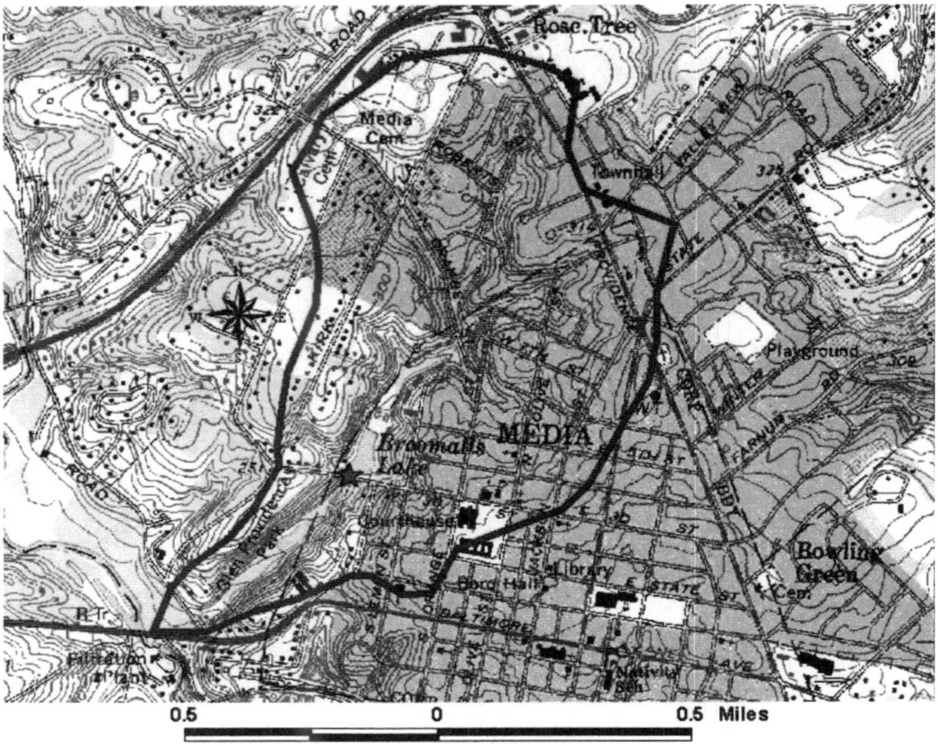

0.5 0 0.5 Miles

★ Third Street Dam

FIGURE 7-2 A typical topographical map of a watershed
Source: Pennsylvania Spatial Data Access web site: www.pasda/psu.edu/access/quad.

The watershed description includes both a set of maps, typically GIS based, and a narrative description. The base map (a USGS 1:24,000 map or other similarly detailed map) has the following overlays:

Base map	The watershed boundary
Overlay 1	Identification of points of water withdrawal and an outline of aquifers that have the potential to yield more than 100,000 gpd (380,000 L/d)
Overlay 2	Areas that have high erosion potential and information on animal populations

Overlay 3	Identification of any state groundwater discharges having state or federal permits, all surface water discharges having National Pollution Discharge Elimination System (NPDES) permits, and solid waste facilities
Overlay 4	An outline of all sewered areas
Overlay 5	Location of all gasoline stations and petroleum storage facilities and pipelines
Overlay 6	Floodplains, wetlands, streams, and water bodies
Overlay 7	Generalized land use, such as • Residential areas • Industrial areas • Commercial areas • Mixed urban areas • Croplands, orchards, golf courses • Animal-rearing areas • Other agriculture • Vacant or undeveloped land • Transportation facilities
Overlay 8	Hazardous-waste sites
Overlay 9	Water quality monitoring sites
Overlay 10	Land ownership • Open space owned or under the control of the water supplier or state, county, or municipal government • Lands in conservation restrictions in perpetuity • All lands owned or restricted by written agreements or deed restrictions with landowners

An additional overlay consisting of recently taken digital aerial photographs can also be very useful in assessing the relationships among geographic features within the watershed.

The development of a robust watershed protection plan will require appropriate expertise and a significant level of effort. While that may be costly, watershed planning is essential if a utility wishes to receive a waiver for the surface water filtration requirements of the

SDWA. Under the 1996 amendments to the SDWA, the state, with USEPA concurrence, can waive filtration requirements if a watershed is undeveloped, uninhabited, or in consolidated ownership and has access that is controlled, resulting in exceptionally high water quality. If the utility believes that the chances of a filtration waiver are good, then the cost of watershed planning will be small compared with the cost of constructing filtration facilities. A reference to the goals of a state watershed control plan developed to help water sources to qualify for a filtration waiver is included in the section titled Source Water Protection for Filtration Waiver Attainment in this chapter.

Watershed Control Programs

While managing watersheds to maintain or enhance water quality is necessary for utilities that wish to avoid filtration, it should arguably be a goal for all utilities that use surface water. Water utility managers should see to it that all water quality hazards upstream of the utility are eliminated. Any continuous contaminant release within a watershed, though it may appear to be causing little or no damage, can become dangerous in the long run. Table 7-1 lists some of the program elements for selected utility source management programs and illustrates the extensive watershed control program implemented by some water utilities.

Complete control

The ideal situation from the standpoint of water source protection is a completely fenced, utility-owned watershed. Although this is usually not possible, both Seattle, Wash., and Portland, Ore., do have this type of controlled supply.

Utilities that do not own an appreciable percentage of their watershed must develop protection strategies that rely more on the cooperation of the public, the local civil authorities, and the state.

An example of a cooperative approach to source protection is the Portland, Maine, Water District, which serves a relatively small population but has a 450-mi^2 (1,166-km^2) watershed that includes portions of 21 communities (Figure 7-3). The district makes a distinction between water that flows directly to Sebago Lake and water that reaches the lake after passing through other lakes and ponds. The distinction is made because the latter water receives some beneficial treatment in reducing phosphorus and other compounds. Portland's definitions for this distinction are as follows:

- **Direct watershed** consists of all land that drains directly to a lake without passing through another lake or pond.
- **Indirect watershed** of a lake consists of all watersheds of upstream lakes and ponds.

TABLE 7-1 Elements of source water management programs for selected surface systems with highly protected managed watersheds

	Seattle, Wash.	Portland, Ore.	New York City, N.Y.	Boston, Mass.	Bridgeport Hydraulic Company, Conn.	Hackensack Water Company, N.J.
Type of supply	Reservoir, river	Reservoir, river	Reservoir	Reservoir, river	Reservoir	Reservoir
Safe yield, mgd (ML/d)	170 (643)	108 (409)	1,290 (4,883)	300 (1,136)	66 (250)	125 (473)
Filtered	No	No	No	No	Partial	Yes
Watershed area, mi^2 (km^2)	162 (420)	106 (275)	1,928 (4,994)	474 (1,228)	96 (249)	113 (293)
Program Elements						
Watershed owned or controlled, mi^2 (km^2)	156 (404)	106 (275)	138(+) (357)*	180 (466)	26 (67)	10 (26)
%	100	100	7(+)	38	27	9
In situ treatment						
Aeration	—	—	—	—	X	—
Algicide or herbicide	—	—	X	X	X	X
Wildlife control	X	Partial	—	Partial	Partial	—
Forest management	X	X	X	X	X	—
Emergency response	X	X	X	X	X	X
Sanitary survey	X	—	X	X	X	X
Source monitoring	X	X	X	X	X	X
Watershed inspection	X	X	X	X	X	X
Security patrol	X	X	X	X	X	X
Fencing	—	Partial	—	Partial	—	X
Public education	X	X	—	X	X	X

Source: Water Quality and Treatment (1990).

*Approximately one half of watershed dedicated to forest preserves.

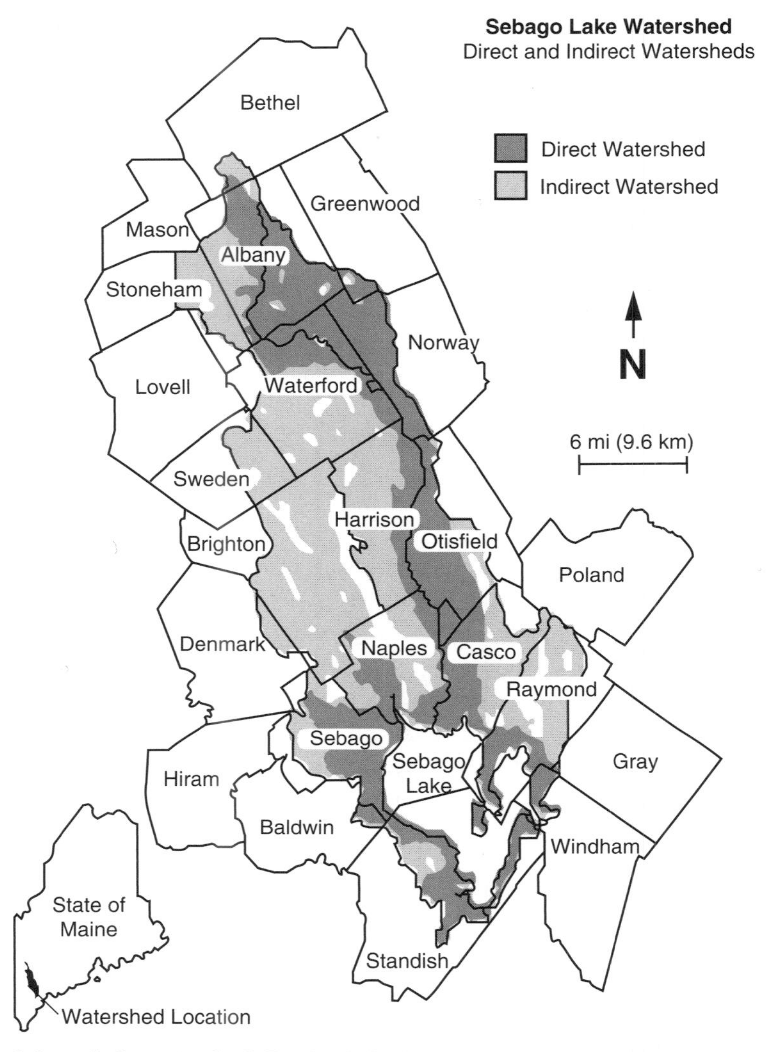

FIGURE 7-3 Sebago Lake watershed, Portland (Maine) Water District, which includes portions of 21 communities
Source: Portland Water District, Portland, Maine.

The implementation of the source protection program calls on portions of 20 federal, state, and local laws. In addition, 10 to 15 sets of rules and regulations that carry the force of law are implemented when necessary. In most cases, when the need arises to invoke a state law, the appropriate state agency takes the lead in approaching the violator. The same is true for local problems—if the utility's watershed inspector finds a malfunctioning septic system, for example, the local plumbing inspector or code enforcement officer is contacted to see that repairs are made.

Local support

Since very few utilities have the police power to enforce source water protection, most must secure protection services from the authorities that do have such policing power. The best way to secure this kind of protection is for the utility to make its case to the public. Most citizens appreciate the importance of protecting local water resources and will be willing to endorse the use of government power and public funding to that end.

To ensure that citizens understand the resource protection issues in their own community, an ongoing public education effort should be sustained as part of a utility's source protection program. Operators should be available to talk to planning and zoning boards, service clubs and other social groups, and schools. Facilities should be maintained so that tours are always welcome and visitors receive the best impression.

Members of the news media can likewise become valuable allies in the cause of watershed protection. The utility will likely need to take the initiative to get supportive coverage from local newspaper, radio, and television. If the utility does not have someone on the staff capable of doing this, a public relations firm can be hired to foster a cooperative relationship between the water company and the media and, ultimately, the public.

Watershed protective districts

Special watershed protective districts are another mechanism that has proven very effective in protecting water sources and watersheds. They are usually created when the geographical area to be protected is quite large and there are a variety of interests that will benefit from environmental protection.

One of the best known of these districts is the one that was organized to protect Lake Tahoe. Because the lake has its shoreline in two states (California and Nevada), protective legislation has been passed by Congress rather than by the states independently.

In Maine, the Saco River Corridor Commission was founded by landowners for their own protection. However, the utility that uses the river as a water supply also benefits from the controls that are in place. The river extends a fairly long distance, which would be beyond the capabilities of the utility to protect if it had to act alone. However, the landowners have the power of eminent domain and the authority to level assessments. It can be difficult to form protective districts because their scope can be quite broad and some interests do not see them as beneficial. In many cases, a water source has to be obviously damaged or severely threatened before diverse groups perceive the threat and are willing to cooperate.

The creation of these districts can help a utility by giving added protection to the watershed and water body. However, since the utility does not have direct control over watershed activities, it will need to be deliberate in making its needs and interests known as one member of a board of directors whose members represents other interests as well.

Land Management

Beyond the social and political aspects of watershed management, the actual practices of landowners and land users must be monitored and managed.

Forest management

A forest is a familiar and valuable feature of many watersheds in the United States. Forests protect water quality, and many utilities manage the forests as a source of income. Timber harvesting, if not done carefully, can be extremely damaging to a watershed and its water quality. Consequently, it is important to require that environmentally responsible tree cutters who are experienced in best management practices (BMPs) for forestry are responsible for harvesting the timber.

If current practices do not adequately protect the water source, a registered forester should be consulted for guidance in identifying better approaches. It might be advisable to implement innovative solutions to ensure that the watershed is adequately protected. One example of going beyond conventional practices is to require that horses be used for tree-cutting work because motor-powered machinery is too damaging to the forest floor.

Likewise, home and commercial land development must be undertaken responsibly. If a project requires that large areas must be stripped of vegetation, extreme care must be exercised to make sure that erosion of the bare land does not damage the source water. Holding ponds and sediment basins that are constructed to minimize damage should be left in place after the project is completed so that the benefits they offer can continue. Revegetation of the watershed with woody species should be encouraged to stabilize soil and so that infiltration of water to recharge groundwater will resume.

Enforcing the construction runoff requirements of the NPDES can help to minimize the immediate adverse effects of construction projects. The permanent removal of vegetation and the increase in the amount of impervious area, however, will have a long-term impact.

Land erosion

Erosion of watershed land adjacent to the shores of streams and lakes is a significant water quality concern. Erosion can result from footpaths and bike paths, but it can be much more serious with motor vehicles. Automobiles and trucks are an obvious cause for erosion, yet the most serious threat comes from motorbikes, motorcycles, and all-terrain vehicles. Both on- and off-trail uses damage vegetative cover and compress the soil surface. Overcompaction of soils, especially on slopes, ultimately results in serious soil losses during precipitation events when overland or sheet flow takes place. The use of off-road vehicles near shorelines of streams and lakes should therefore be discouraged, as should the overuse of paths that are unprotected by gravel or wood chips.

Finally, loss of soil due to erosion in stream channels results in transport of heavy sediment loads into a drinking water reservoir. Heavy sediment loads decrease water quality and, because the sediments eventually settle, reduce the reservoir's capacity. Revegetation of stream banks and contiguous areas damaged by recreational activities should be a priority management effort in a surface drinking water watershed. In addition, limiting high flow velocity throughout the stream channel network would lessen the impacts caused by soil erosion.

Management of agricultural practices

A source water protection program is also concerned with agricultural activities in a watershed. Farmers are usually in contact with conservation districts or extension agents (government officials who serve as consultants to agriculture), and the utility can enlist the aid of these experts to act as a liaison with the agricultural community. These experts can assist by encouraging agricultural BMPs that minimize the movement of soil, herbicides, and pesticides into nearby surface water bodies.

Management in urban areas

Homeowners close to a water body should be informed of the harmful effects of lawn and gardening practices that are improperly conducted. The use of nitrogen and phosphorus fertilizers to turn lawns green can also turn lakes green with algae. The following list of recommendations was taken from one of the first Source Water Assessments provided to the Philadelphia Suburban Water Company through a grant from the Pennsylvania Department of Environmental Protection (Schnabel Engineering 2001). Zone A is the portion of the watershed that could transport a contaminant to the water supply intake within a five-hour period after introduction:

- Cooperative action to label and protect storm drains and tributary systems in Zone A, including inventorying and labeling storm and roadway drains and road crossings.
- Delineation and identification of the boundary of Zone A on township and borough zoning maps.
- Enactment of ordinances that require measures to infiltrate or retain and treat stormwater runoff to remove sediment and nutrients and capture potential spills (particularly in Zone A); as well as ordinances that might encourage retrofitting of existing storm drains and road drains to protect source water, including measures to protect and restore forested riparian buffers.
- Grant funds or loan programs to commercial/industrial/residential areas to pay for improvements to stormwater management facilities.

- Retrofitting by the Department of Transportation of drainage systems from state highways that drain directly into tributaries of Zone A.
- Improved management of septic systems.
- Improved monitoring and/or instrumentation of sewer collector systems and lift stations, particularly in Zone A.
- Adoption and implementation of Act 537 wastewater facilities plans to accommodate future growth while addressing issues of failing septic systems and to preserve (and if possible, enhance) quality and quantity of water in both Zone A and Zone B.
- Cooperative action to control Canada geese and other waterfowl on the watershed, particularly within Zone A.
- Cooperative streambank and forested riparian corridor restoration projects.

Management of transportation facilities

If highways or railroads run through or close to water bodies used to supply raw water to the utility, the utility has a stake in ensuring that local emergency response personnel, including police and fire departments, know the proper procedure for dealing with spills. Flushing spilled material off a highway to prevent fire or traffic disruption could result in a water supply crisis. For example, if a fire department uses water to extinguish a fire in an agricultural supply warehouse, it will probably flush pesticides and other chemicals into a nearby stream. Depending on the particular circumstance, a better approach might be to use nontoxic fire retardants or mist the surrounding area to prevent the fire from spreading, and then allow the fire to burn itself out.

Recreational Use of Lakes and Reservoirs

Recreational use of water supply reservoirs or lakes has long been a subject of controversy. Studies have shown that controlled recreational use has minimal impact. However, water system managers generally have an aversion to the public being given free rein on water bodies that are used as drinking water sources because of the utility's liability if anything goes wrong.

Because of population pressures from urban areas and the desirability of outdoor activities (especially those involving water), it is difficult to deny the public some use of watershed land. The water resource protection goal then becomes one of managing recreational uses so that they do not threaten water quality.

Levels of recreational use

Recreational use in and around a water body can be constrained at different levels as follows:

1. The least impact on water quality occurs when recreation can be limited to bike and hiking paths, nature walks, and picnic areas. Even with these activities, it is necessary to provide adequate, clean, attractive toilet facilities and a plentiful supply of waste disposal containers that are serviced on a regular basis.

2. The activities just named, plus fishing from shores and boats, comprise the next level of recreational activity. Boats should have no motors or else small, electrically powered motors. Launch sites should be as far from intake structures as possible, and restricted zones should be designated around intakes and marked with buoys.
3. The next level is to have less restriction on fishing and to allow boats with greater horsepower for nonfishing recreation. Speeds should be controlled to prevent erosion from wakes and to promote safety. Again, restricted areas should be enforced. Waste tanks on boats should be sealed and wastes pumped out at an approved facility. Onshore sanitary facilities for people and boat holding-tank flushing stations should be provided. Fuel for boats should be kept as secure and free from spillage as possible.
4. Bodily contact with the water is the last level in this progression. Public swimming areas should be kept as far as possible from intake structures. Good onshore sanitary facilities should also be provided.

In all of these scenarios, water quality monitoring and sanitary inspection are a necessity. It is obvious that as human contact increases, the monitoring requirements become more stringent. If serious bacterial degradation occurs, regulators should be asked to help restrict activities. Where full recreational use has been made available, it becomes politically difficult for a utility to reduce recreational activities to a lower level.

Recreational fees

A generally accepted principal of recreational area management is that the public is more likely to follow the rules if they are charged for the use. Fees can also make the management of recreation on watersheds and water bodies less costly and thus more readily supported by taxpayers or utility customers. Care has to be taken, however, not to raise fees so high that they become a burden.

Other concerns

Adherence to sound watershed management practices by civil and public authorities as well as vigilance on the part of the utility will help ensure high raw water quality at the utility's intake. The discussion of recreation applies only to sources of raw water. Recreation on finished-water reservoirs is universally discouraged by state health authorities and should be completely prohibited.

Source Water Protection for Filtration Waiver Attainment

A well-defined watershed control plan is mandatory if a utility is to receive a waiver for filtration as required by the Surface Water Treatment Rule. Although filtration waivers are part of the national SDWA, states that have assumed primacy have considerable flexibility in determining how the provisions will be enforced. The following information related to avoiding surface water filtration is part of the Massachusetts Watershed Protection Plan (2008). Other states have similar plans. The plan's policy goals are as follows:

1. Provide guidance to water suppliers for preparing and submitting a report that describes a watershed control plan that will meet the USEPA's SDWA requirements to avoid mandatory filtration and to serve as a watershed protection plan.
2. Provide specific and consistent guidance for the protection, preservation, and improvement of raw water quality of surface water supplies. (These sources are open to the atmosphere and subject to surface runoff, or they may be groundwaters that originate from a surface water supply.)
3. Provide a comprehensive approach to identify and control existing and potential sources of pollution of surface water supplies.
4. Link land use activities to water quality concerns, and ensure that a consistent set of standards and safeguards are applied for reducing vulnerability to contamination from inappropriate land uses.
5. Encourage cooperation among the water supplier, landowners, and local authorities sharing a resource area to set up agreements to control discharges to the environment and to respond to emergencies.
6. Provide for an ongoing monitoring program that will be used by the state Department of Environmental Protection to determine the effectiveness of the watershed controls on raw water quality.

GROUNDWATER PROTECTION

Groundwater protection can be instituted at federal, state, and local levels. Regulatory programs typically require permits for well installation and may involve the regulation of land uses near well fields. Figure 7-4 illustrates some of the many potential sources of groundwater contamination that regulatory programs can help address. One example of an interstate authority with a major focus on groundwater quantity/quality is the Delaware River Basin Commission (DRBC). The DRBC has implemented regulations for groundwater-protected area in its region of concern, including in parts of Pennsylvania, New York, and Delaware. See the DRBC website for more information (//www.state.nj.us/drbc).

Aquifer delineation

Groundwater comes from an aquifer that is recharged by percolating and infiltrating surface water. The only way to tell what surface water is recharging a particular well's aquifer is to define the land area over the portion of the aquifer used by the well. This land area is called the aquifer recharge area or the wellhead protection area (WHPA).

In the past, many states used a fixed radius around a well to determine the protected area. The radius was commonly 100 ft (30 m) for a private well and as little as 300 ft (90 m) for a public water supply well. Utilities were sometimes required to own the land within that circle or at least to have control over its use.

TABLE 7-2 Susceptibility analysis for the Crum Creek Watershed

Activity	Matrix A Potential for Contamination	Matrix B	Matrix C Potential Impact	Matrix D Susceptibility Rating
Spills in Zone A	**High**	**High**	**High**	**A**
	Short travel time, high persistence	High quantity, high A result	High sensitivity, high B result	No control practices, high C result
Spills in Zone B	**High**	**High**	**High**	**A**
	Medium travel time, high persistence	High quantity, high A result	High sensitivity, high B result	No control practices, high C result
1. 33-in. sewage pipe	**High**	**High**	**High**	**D**
	Short travel time, medium persistence	Medium quantity, high A result	High sensitivity, high B result	Regulated containment, high C result
2. Waterfowl	**Medium**	**Low**	**Medium**	**B**
	Short travel time, low persistence	Low quantity—domestic level, medium A result	High sensitivity, low B result	No control practices, medium C result
3. Development	**Medium**	**Low**	**Medium**	**D**
	Short travel time, low persistence	Low quantity—domestic level, medium A result	High sensitivity, low B result	Nonpoint source with best management practices, medium C result
4. Sewage lift station	**Medium**	**Medium**	**High**	**D**
	Short travel time, low persistence	Medium quantity, medium A result	High sensitivity, medium B result	Regulated containment, high C result
5. Sewage lift station	**Medium**	**Medium**	**High**	**D**
	Short travel time, low persistence	Medium quantity, medium A result	High sensitivity, medium B result	Regulated containment, high C result

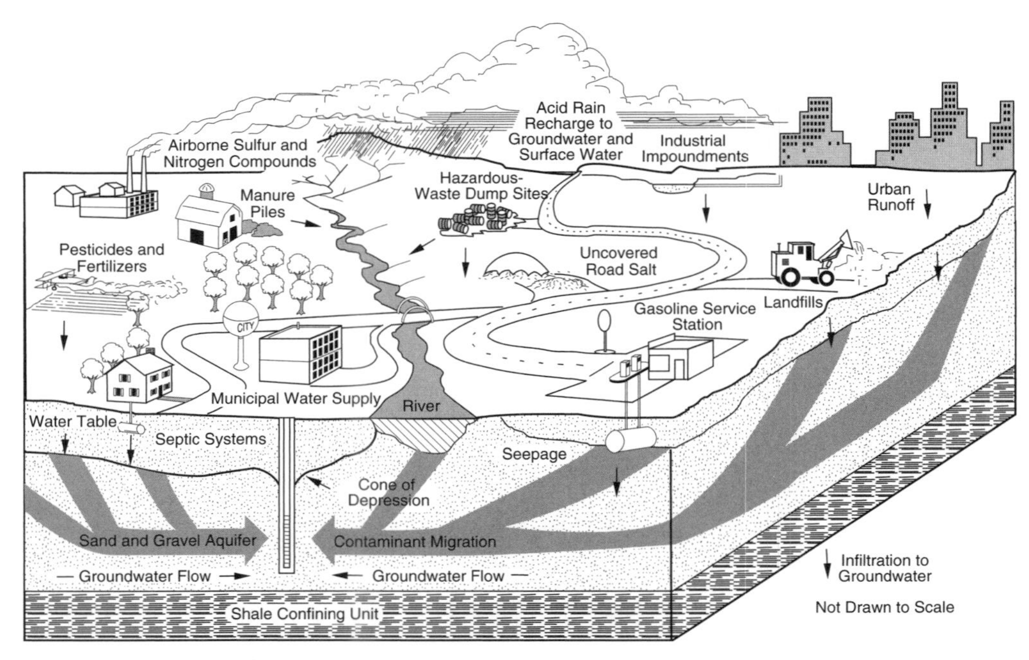

FIGURE 7-4 Sources of groundwater contamination
Source: Wellhead Protection (1987).

If a well is relatively deep and properly constructed, providing protection out to a fixed radius around the well is probably adequate in most cases for keeping microbiological contaminant levels sufficiently low. However, this fixed radius has little bearing on contamination from chemicals. Some chemicals travel up to thousands of feet (meters) within an aquifer. In other words, depending on local geological formations, a relatively small area or a very large area may be required from the WHPA.

If the region has already had its groundwater mapped by the USGS or a state geology department, the scientific information needed to delineate the recharge area for a well may be readily available. Unless the area has very easily determined borders, it is best for a utility to hire a firm that has expertise in hydrogeology and can provide mapping of aquifers for further characterization.

Sources of Contamination

When the WHPA or watershed boundary for a water source has been determined, the next step is to develop a complete inventory of the potential contamination sources within the area. This does not have to be an expensive project, and it does not have to be done by water utility personnel. It could become a public service project; local volunteers from scouting organizations, senior citizens' groups, service clubs, environmental organizations, and volunteer fire department groups have proven very effective in doing this job for several utilities.

Where volunteers are involved, each volunteer or team of volunteers is furnished with a map, or a portion of a map, and asked to mark potential contamination sources on it. A list of common sources (Table 7-3) should be furnished to the volunteers. Detailed information on locations of particular concern can be compiled by the volunteers or followed up by more experienced personnel.

Local and area zoning maps and master plans should also be reviewed to identify possible future sources of contamination from industrial parks, pipelines, sewage treatment plants, waste disposal sites, and other similar developments. In addition, state environmental agencies can be contacted regarding their records of the locations of underground storage tanks or information about other permitted discharges to surface waters.

Once all the data have been collected, the threats to source water quality can be ranked. Table 7-2 is an excerpt from a susceptibility analysis conducted during a source water assessment for a reservoir in Pennsylvania. Figure 7-5 contains an overview and matrix definitions for this assessment.

A potential source of concern for both groundwater and surface water sources is how the contamination of one affects the quality of the other. Contaminated surface water entering the phreatic zone (the saturated, subsurface zone), for example, will eventually mix with and contaminate the groundwater below. In addition, contaminated groundwater discharging to wetlands and streams can degrade their quality. Although the interaction varies from simple to complex, it is important to take into account the connections between groundwater and surface water.

Sole-source aquifers

One way to protect an aquifer is to have it designated as a sole-source aquifer under federal regulations. Under provisions of the SDWA, the USEPA can, on its own initiative or as the result of a petition, designate an aquifer as a sole source of water for a geographic area. After that determination has been published, no federal money may be used or granted for any purpose that could result in the contamination of that aquifer.

Sole-source protection has been granted to aquifers as small as those located on rocky islands off the coast of Maine and as large as the 175-mi-long (282-km-long) Edwards Aquifer that supplies the San Antonio, Texas, area. The Edwards Aquifer was the first to receive sole-source designation under the SDWA. A map showing locations of sole-source aquifers can be obtained from USEPA's Office of Ground Water and Drinking Water or at its website (www.epa.gov/safewater/swp/sumssa.html).

TABLE 7-3 Common sources of groundwater contamination

Category	Contaminant Source	
Agricultural	Animal burial areas	Irrigation sites
	Animal feedlots	Manure-spreading areas or pits
	Fertilizer storage or use	Pesticide storage or use
Commercial	Airports	Jewelry or metal plating facilities
	Auto repair shops	Laundromats
	Boatyards	Medical institutions
	Construction areas	Paint shops
	Car washes	Photography establishments
	Cemeteries	Railroad tracks and yards
	Dry cleaners	Research laboratories
	Gas stations	Scrapyards and junkyards
	Golf courses	Storage tanks
Industrial	Asphalt plants	Petroleum production or storage
	Chemical manufacture or storage	Pipelines
	Electronics manufacture	Septic lagoons and sludge
	Electroplaters	Storage tanks
	Foundries or metal fabricators	Toxic and hazardous spills
	Machine or metalworking shops	Wells (operating or abandoned)
	Mining and mine drainage	Wood-preserving facilities
Residential	Fuel oil	Septic systems or cesspools
	Furniture stripping or refinishing substances	Sewer lines
	Household hazardous products	Swimming pools (chemicals)
	Household lawns	
Other	Hazardous-waste landfills	Recycling or reduction facilities
	Municipal incinerators	Road-deicing operations
	Municipal landfills	Road maintenance depots
	Municipal sewer lines	Stormwater drains or basins
	Open burning sites	Transfer stations

POTENTIAL FOR CONTAMINATION

MATRIX A — Time of Travel vs. Persistence

Time of Travel	Short	Medium	Long
Surface Water Intakes	Zone A = 5 hours	Zone B = 25 hours	Zone C = remainder of watershed

Persistence	High	Medium	Low
Contaminant ability to move in the environment	If contaminant has been known to contaminate water with concentration greater than the MCL or in significant concentration.	Significant concentrations contaminate water, but may be partially removed by geologic materials.	If area's geology allows for removal of the contaminant

MATRIX B — Matrix A vs. Quantity

Quantity	High	Medium	Low
	Clearly associated with commercial- or industrial-sized operations with a minimum 10x a reportable release, event, or equivalent	Reportable releases, events, up to 10x those quantities associated with commercial- or industrial-sized operations	Clearly on a domestic level

POTENTIAL IMPACT

MATRIX C — Potential for Contamination (Result of Matrix B) vs. Sensitivity

Sensitivity	High	Medium	Low
Sensitivity of drinking water to contaminants	Surface water sources—short travel times, limited processes for mitigation (Zone A)	Longer travel times, some ability to mitigate (Zone B)	Overlying geologic materials are expected to provide some treatment for infiltrating water (Zone C)

SUSCEPTIBILITY RATING

MATRIX D — Potential for Release (from Potential for Release Table) vs. Potential Impact (Result of Matrix C)

Methodology taken from the Commonwealth of Pennsylvania's Source Water Assessment and Protection Program document (document 383-5000-001 at the following Web site: http://www.dep.state.pa.us/dep/subject/All_Final_Technical_guidance/bwsch/bwsch.htm)

FIGURE 7-5 Susceptibility analysis of drinking water sources to contamination definitions of matrix terms

Source: Pennsylvania Dept. of Environmental Protection (2000).

Other environmental regulations

There are a number of other national laws that protect aquifers, including:

- The Resource Conservation and Recovery Act (RCRA), which regulates the storage, transportation, treatment, and disposal of solid and hazardous wastes. It is principally designed to prevent contaminants from leaching into groundwater from municipal landfills, underground storage tanks, surface impoundments, and hazardous-waste disposal facilities.
- The Comprehensive Environmental Response, Compensation, and Liability Act (Superfund), which authorizes the government to clean up contamination caused by chemical spills or hazardous-waste sites that could (or already do) pose threats to the environment. It also allows citizens to sue violators of the law and establishes community right-to-know programs.
- The Federal Insecticide, Fungicide, and Rodenticide Act (FIFRA), which authorizes the USEPA to control the availability of pesticides that have the ability to leach into groundwater.
- The Toxic Substances Control Act (TSCA), which authorizes the USEPA to control the manufacture, use, storage, distribution, or disposal of toxic chemicals that have the potential to leach into groundwater.
- The Clean Water Act, which authorizes the USEPA to make grants to the states for the development of groundwater protection strategies and authorizes several programs to prevent water pollution.
- The Safe Drinking Water Act, which also requires the USEPA to regulate underground disposal of wastes in deep wells.

Each of these national laws has been designed for implementation by the states in cooperation with local government. A major reason for emphasizing local action is that groundwater protection generally involves specific decisions concerning land use, and most land use controls are implemented by local authorities who have been given regulatory power through state statutes or regional agreements.

State-Level Aquifer Protection

All states have passed some type of legislation pertaining to the protection of aquifers. This legislation falls into a number of legal subject areas. Some states have minimal programs, whereas others have very comprehensive ones. State water utility associations are generally active in promoting protective legislation.

Typical features of state programs are as follows:

- **Comprehensive planning** is an activity that many states have required for local jurisdictions. Cities, towns, counties, and local districts have been mandated to develop plans that protect a state's groundwater.

- **Groundwater classification** involves reviewing all groundwater sources and categorizing them so that appropriate types of protection can be determined and implemented.
- **Standard setting** involves the evaluation of groundwater to determine the degree of contamination. It can also include required remedial action.
- **Land use management** determines whether planned land uses may cause contamination and regulates any such uses.
- **Funding proposals** are used to secure funds for contaminated water cleanup and for supplying alternative drinking water in an emergency.
- **Agricultural chemical regulations** are imposed so that pesticides, herbicides, and fertilizers are used judiciously.
- **Underground storage tank regulations** are imposed to control possible leakage of hazardous material that is generally stored underground. The regulations establish criteria for the registration, construction, installation, monitoring, repair, abandonment, and financial liability related to underground storage.
- **Water usage control** is exercised to allocate the use of water, with the goal of preventing saltwater intrusions or other overutilization problems.

State programs are the most valuable general aids to the water utility. All water utility managers should ensure that there is adequate funding of state programs to protect all waters of their state.

Local Wellhead Protection Programs

In addition to the resources available to the utility through national and state authorities, a wellhead protection program (WHPP) adds control at the local level. Ideally, water utilities will take the lead in developing the WHPP that involves their own groundwater sources to ensure that the WHPP is implemented in the best interest of the utility. The 1986 amendments to the SDWA require that each state develop a comprehensive WHPP, but that program largely depends on active participation at the local level. More information about the WHPP and its implementation in different states can be found at the web site of the USEPA.

A WHPP requires cooperation between the local civil authority and the water utility. The most important technical step is determination of the recharge area, as discussed previously. It is likely that the utility will have to use its own budget for funding this effort.

Determination of the wellhead protection area

There are different ways to delineate an area that needs protection, and engineering or geological consultants have their own preferences as to which method does the best job. The state health, environmental, or geological agency that is responsible for WHPPs can identify preferred practices. Private consulting firms provide services that use techniques ranging from the less accurate (but also less expensive) fixed-radius methods to the more accurate (but also more expensive) analytical or numerical models. Details and examples of delineation methods are available from the USEPA (see end notes for this chapter). Figure 7-6 illustrates a wellhead protection area laid out on a topographic map.

Numerical modeling can be highly useful for some situations. Based on parameters that include hydraulic conductivity, specific yield, and streambed permeability, a computer model can be developed to show the direction and rate of subsurface flow relative to the location of the wellhead. This model can then be verified and modified after comparison with known events. The modeling stages can be characterized as follows:

1. Select the study area and model grid
2. Prepare the data
3. Calibrate the model
4. Verify the model
5. Check the model sensitivity

Once a wellhead protection area has been delineated, a determination should be made of the time of travel for percolating water to reach the well. The utility should be more concerned about possible pollution in a zone that has 30 days' travel time than one that has 500 days. It is often useful to test the model using a range of assumptions to get a range of outputs that includes a worse-case scenario.

Conducting a contaminant source inventory

The second step in establishing a WHPP is to conduct a contaminant source inventory. Contaminant sources can be identified by examining available compilations of businesses, industries, and known contamination sites. Much of these data may be available on computerized maps or in a digital form that can be readily imported into a geographic information system. Addresses in a spreadsheet, for example, can be readily mapped to show whether a given addresses lies in or near the WHPA. After the inventory is compiled and mapped, volunteers or trained personnel can conduct fieldwork to obtain more specifics on particular contaminant sources.

Wellhead protection area management

The WHPA, the contaminant sources, and future development within the WHPA must then be managed. The implementation of zoning restrictions can be of immense help in doing this. Citizens, authorities, and business and industrial leaders should be informed of the importance of protecting the water supply and then asked to enact zoning laws that will control land use. Management of existing hazards can best be handled through personal contact and public education.

Installation of monitoring wells may also be necessary to ensure that major industries are not causing groundwater pollution. In some cases, industries have even been willing to install the monitoring wells and pay for the sampling.

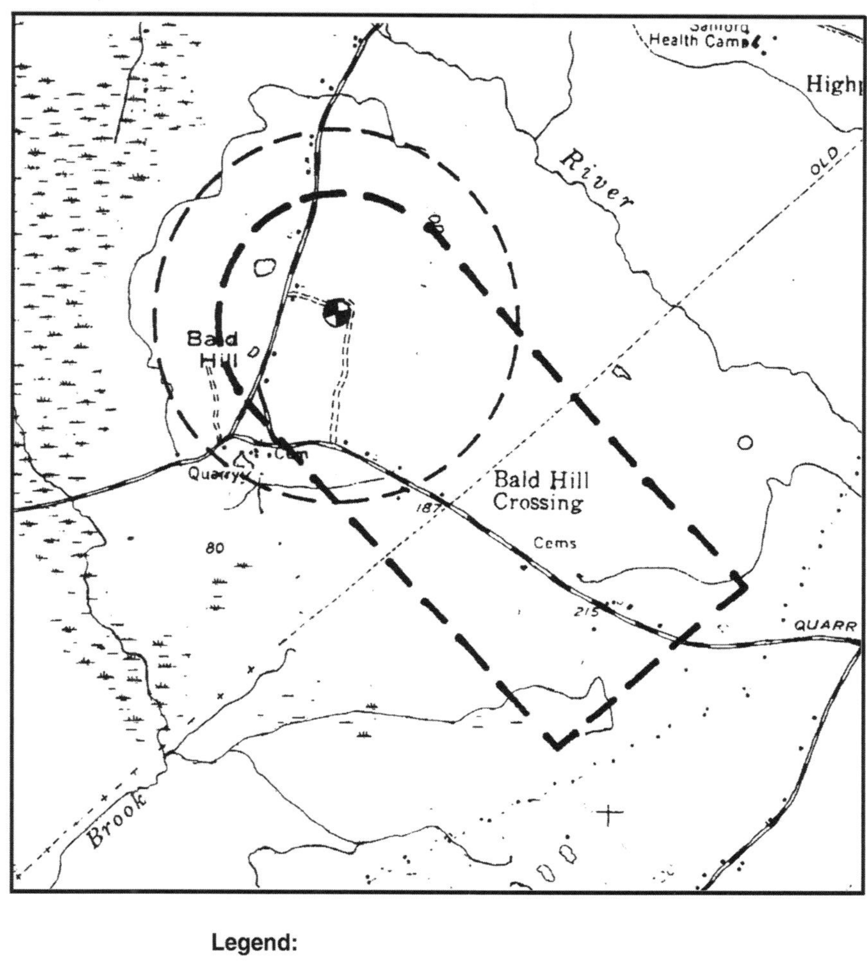

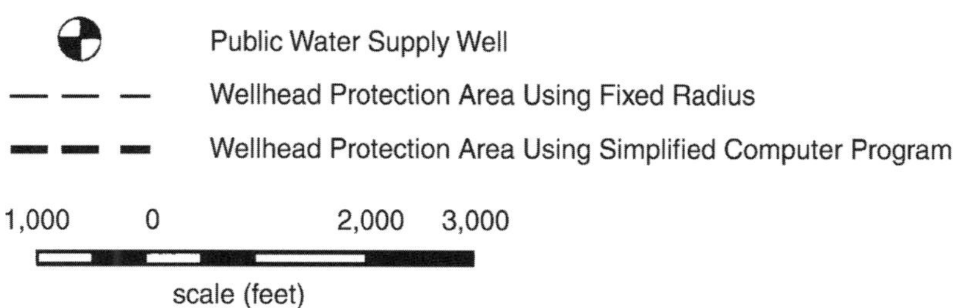

FIGURE 7-6 A wellhead protection area laid out on a topographic map
Source: Protecting Local Ground-Water Supplies Through Wellhead Protection (1991).

Contingency plan

A WHPP is not complete until a contingency plan is prepared, detailing the steps to be taken if contamination ever occurs. The plan should provide for an alternative water supply for both short-term and long-term emergencies.

Continuing review

Finally, the WHPP should be an active program. Built-in review periods with short intervals for minor review and longer periods for major review should be written into the plan.

Sanitary Surveys

A periodically conducted sanitary survey is one of the most important water quality control measures that any water utility can implement to determine how well it is fulfilling its mission. A sanitary survey is an on-site review of the water source, facilities, equipment, operation, and maintenance of the public water system. A comprehensive survey involves the evaluation of the adequacy of all aspects of producing and distributing drinking water.

A sanitary survey of each water source is necessary for proper water source management. The utility may conduct it by following a standard form supplied by the state health authority. The form should be designed to identify areas of concern throughout the supply system, including the entire watershed or aquifer recharge area. A qualified person, preferably a sanitary engineer or someone having equivalent training, should complete this portion of the form. The survey should determine the potential for contamination of the supply and encompass all installations, activities, and contamination sources.

To ensure that changes in water quality within the system can be spotted, the survey should include water sampling at specific points considered to be representative of the quality in each part of the system. Some portions of the system should be surveyed annually; others can be evaluated at less frequent intervals. Important changes in water quality found in annual sampling should be investigated. If the reason for the change cannot readily be determined, a complete survey should be performed.

Vandalism and Terrorism

September 11, 2001, has dramatically affected the level of concern and the procedures undertaken to ensure the security of water supplies in the United States. Concern over high-hazard dams, water treatment facilities, and inlet and outlet pipes led to the development of various vulnerability assessment methodologies. Although water utility personnel can be thankful that no major violence or violent acts have been associated with water sources or a water supply to any important extent, security has become a priority. Even in countries at war or suffering from civil unrest, water supplies have not yet been used as agents to carry toxic or disease-causing materials. However, the possibility of vandalism or terrorism aimed at a water system must be seriously considered, and dealing with it should be part of any utility emergency plan.

Utilities can take some comfort in the fact that it is very difficult for someone to effectively poison the water supply. If water were delivered to customers in small volumes to be used exclusively for drinking, it would be relatively easy for a terrorist to add enough dangerous material to cause human illness. However, the fact that drinking and cooking account for only a small portion of the water delivered to customers by a public water system is an important safety factor. It means that a rather large volume of poison would have to be introduced to the water supply to create a high enough concentration to cause sickness from drinking the water.

A more credible threat is the sabotage of treatment and pumping facilities. Arson or use of explosives could easily prevent the delivery of water to at least a portion of the transmission and distribution system for most water utilities. Prevention of those acts is practical only in the face of a confirmed threat, but responding to the resulting damage should be considered a part of every utility's emergency plans. Fires and explosions are much more likely to happen as a consequence of human error, or an accident, than of malice.

Following September 11, 2001, in response to heightened concerns about the security of water systems, the USEPA and the US Department of Energy (Sandia Laboratories) have taken several actions to help protect the nation's drinking water, wastewater infrastructure, transmission lines, and dams. These actions include the following:

- **Alerts and notices** are sent to all utilities and local law enforcement outlining security measures, resources available, and advice on monitoring and treatment.
- **Vulnerability assessments and remediation plan templates** for drinking water utilities are now available.
- **Model emergency operations plans** for both wastewater and drinking water systems are also available.
- **General security overview training**, developed through partnership with American Water Works Association, covers a broad spectrum of security issues.
- **Vulnerability assessment overview and implementation training** are the subjects of detailed presentations and workshops on what constitutes an effective vulnerability assessment for drinking water utilities and how they can be applied.
- **Information sharing** will be improved because of the development of a secure Information Sharing and Analysis Center, which partners the Association of Metropolitan Water Agencies, the FBI, and the USEPA.

SELECTED SUPPLEMENTARY READINGS

Black, P.E. 1996. *Watershed Hydrology,* 2nd ed. Chelsea, Mich.: Sleeping Bear Press.

Citizens' Guide to Ground-Water Protection. 1990. Washington, D.C.: US Environmental Protection Agency, Office of Drinking Water.

Cooke, G.D., and R.E. Carlson. *Reservoir Management for Water Quality and THM Precursor Control.* 1989. Denver, Colo.: Awwa Research Foundation and American Water Works Association.

Drinking Water Quality Enhancement Through Source Protection. 1977. Ann Arbor, Mich.: Ann Arbor Science.

Dunne, T., and L.B. Leopold. 1978. *Water in Environmental Planning.* San Francisco: W.H. Freeman and Company.

Guide for Conducting Contaminant Source Inventories for Public Drinking Water Supplies. 1991. Washington, D.C.: US Environmental Protection Agency, Office of Drinking Water.

Local Financing for Wellhead Protection. 1989. Washington, D.C.: US Environmental Protection Agency, Office of Drinking Water.

Massachusetts Watershed Protection Plan. 2008. Boston, Mass.: Department of Environmental Protection.

Pennsylvania Department of Environmental Protection. 2000. *Source Water Assessment and Protection Program.* Document No. 383-5000-001. Harrisburg: Pennsylvania Department of Environmental Protection.

Protecting Local Ground-Water Supplies Through Wellhead Protection. 1991. Washington, D.C.: US Environmental Protection Agency, Office of Drinking Water.

Robbins, R.W., J.L. Glicker, D.M. Bloem, and B.M. Niss. 1991. *Effective Watershed Management for Surface Water Supplies.* Denver, Colo.: Awwa Research Foundation and American Water Works Association.

Schnabel Engineering Associates, Inc. 2001. *Source Water Assessment, Public Summary Crum Creek Basin—Philadelphia Suburban Water Company, October 2001.* Unpublished public document, Schnabel Engineering Associates, West Chester, Pa.

Sham, C.H., R.W. Gullick, S.C. Long, and P.P. Kenel. 2010. *Operational Guide to AWWA Standard G300, Source Water Protection.* Denver, Colo.: American Water Works Association.

Stewart, J.C. 1990. *Drinking Water Hazards.* Hiram, Ohio: Envirographics.

Symons, J.M., M.E. Whitworth, P.B. Bedient, and J.G. Haughton. 1989. Managing an Urban Watershed. *Jour. AWWA*, 81(8):30.

Water Quality and Treatment, 5th ed. 1999. New York: McGraw-Hill and American Water Works Association (available from AWWA).

Wellhead Protection—A Decision Makers' Guide. 1987. Washington, D.C.: US Environmental Protection Agency, Office of Ground Water Protection.

Winter, T.C., J.W. Harvey, O.L. Franke, and W.M. Alley. 1998. *Ground Water and Surface Water: A Single Resource.* Circular 1139. Denver, Colo.: US Geological Survey.

Glossary

absolute ownership A water rights term referring to water that is completely owned by one person.

absolute right A water right that cannot be abridged.

acid rain Rain or snow that has a low pH because of atmospheric contamination.

acidic water Water having a pH less than 7.0.

adsorption A physical process in which molecules adhere to a substance because of electrical charges.

advanced wastewater treatment The treatment of sewage beyond the ordinary level of treatment.

aeration The process of bringing water and air into close contact to remove or modify constituents in the water.

aesthetic Of, or pertaining to, the senses.

air-stripping Removal of a substance from water by means of air.

algae Primitive plants (one- or many-celled) that usually live in water and are capable of obtaining their food by photosynthesis.

alkaline water Water having a pH greater than 7.0. Also called *basic water*.

alluvial Referring to a type of soil, mostly sand and gravel, deposited by flowing water.

anaerobic Occurring in the absence of air or free oxygen.

annual average daily flow The average of the daily flows for a 12-month period. It may be determined by means of dividing the total volume of water for the year by 365, the number of days in the year.

annular space The space between the outside of a well casing and the drilled hole.

appropriation doctrine A water rights doctrine in which the first user has the right to water before subsequent users.

appropriation–permit system A water use system in which permits to use water are regulated so that overdraft cannot occur.

appropriative rights Water rights acquired by means of diverting and putting the water to beneficial use following procedures established by state statutes or courts.

aquatic life All forms of animal and plant life that live in water.

aqueduct A conduit, usually of considerable size, used to convey water.

aquifer A porous, water-bearing geologic formation. Generally restricted to materials capable of yielding an appreciable supply of water. Also called *groundwater aquifer.*

aquifer recharge area The land above an aquifer that contributes water to it.

artesian aquifer An aquifer in which the water is confined by both an upper and a lower impermeable layer.

artesian well A well in which water pressure forces water up through a hole in the upper confining, or impermeable, layer of an artesian aquifer. In a flowing artesian well, the water rises to the ground surface and flows out onto the ground.

artificial wetland A wet area created by either damming or rerouting a stream and/or grading down to groundwater, eventually resulting in the growth of wetland plants, and the development of wetland hydrology and hydric soils.

atmospheric deposition The process whereby airborne particles and gases are deposited on the earth's surface.

atom The basic structural unit of matter; the smallest particle of an element that can combine chemically with similar particles of the same or other elements to form molecules of a compound.

average daily flow The sum of all daily flows for a specified time period, divided by the number of daily flows added.

bacteria A group of one-celled microscopic organisms that have no chlorophyll. Usually have spherical, rodlike, or curved shapes. Usually regarded as plants.

basic water See *alkaline water.*

beneficial use A water rights term indicating that the water is being used for good purposes.

biofilm A layer of biological material that covers a surface.

bloom An undesirable large growth of algae or diatoms in a lake or water reservoir.

buffered Able to resist changes in pH.

caisson A large-diameter excavated chamber in which work can be done beneath the surface of the earth or water.

capillary action The degree to which a material or object containing minute openings or passages, when immersed in a liquid, usually water, will draw the surface of the liquid above the water surface.

carcinogenic Capable of causing cancer.

casing (1) The enclosure surrounding a pump impeller, into which are machined the suction and discharge ports. (2) The metal pipe used to line the borehole of a well (also called *well casing*).

catchment area See *drainage basin*.

cement grout A mixture of cement and water used to seal openings in well construction.

chemical oxidation The use of a chemical, such as chlorine or ozone, to remove a contaminant from water or change the contaminant.

chemical precipitation The use of a chemical to cause some contaminant to become insoluble in water.

chlorophyll The green matter in plants or algae.

coagulation A treatment that causes particles in water to adhere to each other to form larger particles.

coliform group A group of bacteria predominantly inhabiting the intestines of humans or animals, but also occasionally found elsewhere. Presence of the bacteria in water is used as an indication of fecal contamination (contamination by human or animal waste).

color A physical characteristic describing the appearance of water that is other than clear. Different from *turbidity*, which is the cloudiness of water.

color units (cu) The unit of measure used to express the color of a water sample.

commercial water uses Use of water in business activities such as motels, shopping centers, gas stations, and laundries.

common-law doctrine Laws that are a part of a large body of civil law for the most part made by judges in court decisions rather than being enacted by a legislative body.

compaction A process that renders soil more dense and often reduces water transfer.

comprehensive planning A process that results in widespread community planning.

computer model A means of defining a system using computer programs.

condensation The process by which a substance changes from the gaseous form to a liquid or solid form. Water that falls as precipitation from the atmosphere has condensed from the vapor (gaseous) state to rain or snow. Dew and frost are also forms of condensation on the surface of the earth or vegetation.

cone of depression The cone-shaped depression in the groundwater level around a well during pumping.

confining bed A layer of material, typically consolidated rock or clay, that has very low permeability and restricts the movement of groundwater into or out of adjacent aquifers.

consecutive systems Two or more water systems that take water from a single providing system.

conservation district A governmental entity that acts to protect soil and water.

contaminant Anything found in water other than hydrogen or oxygen.

contaminant source inventory A record of the activities on a watershed or aquifer recharge area that have a potential to contaminate water.

contamination Any introduction into water of microorganisms, chemicals, wastes, or wastewater in a concentration that makes the water unfit for its intended use.

contingency plan A document that details the action of a water utility under specified adverse conditions.

corrosion Destruction by chemical action.

curbs Precast concrete liners for dug wells.

cyst A resistant form of a living organism.

daily flow The total volume of water (in gallons or liters) passing through a plant during a 24-hour period.

demand factor The ratio of the peak or minimum demand to the average demand.

demand management A conservation term that refers to controlling the users' use of water.

density stratification The formation of layers of water in a reservoir, with the water that is the densest at the bottom and least dense at the surface.

detention time The period for which water is held before some further action takes place.

disinfection The water treatment process that kills disease-causing organisms in water, usually by adding chlorine. Contrast with *sterilization*.

dissolved oxygen (DO) The oxygen dissolved in water, wastewater, or other liquid, usually expressed in milligrams per liter, parts per million, or percent of saturation.

dissolved solids Any material that is dissolved in water and can be recovered by evaporating the water after filtering the suspended material.

divide The line that follows the ridges or summits forming the boundary of a drainage basin (watershed) and that separates one drainage basin from another. Also called *watershed divide.*

drainage basin An area from which surface runoff is carried away by a single drainage system. Also called *catchment area, watershed,* or *watershed drainage area.*

drawdown The difference between the static water level and the pumping water level in a well.

drawdown method A testing procedure to determine the characteristics of an aquifer or well.

dual system A double system of pipelines, one carrying potable water and the other water of lesser quality.

effluent Water flowing out of a structure such as a treatment plant.

electrical conductivity (EC) A test that measures the ability of water to transmit electricity. Electrical conductivity is an indicator of dissolved solids concentration. Normally an EC of 1,000 μmhos per square centimeter indicates a dissolved-solids concentration of 600–700 mg/L.

eminent domain The power to seize property.

ephemeral stream (1) A stream that flows only in direct response to precipitation. Such a stream receives no water from springs and no continued supply from melting snow or other surface source. Its channel is above the water table at all times. (2) Streams or stretches of streams that do not flow continuously during periods of as much as one month.

Escherichia coli (E. coli) A bacterial species; one of the fecal coliforms.

eutrophication A process by which a lake produces too many single-celled algae cells resulting from the input of the nutrients, which cause their excessive growth. A eutrophying water body is usually characterized by a marked decrease in water clarity, oxygen depletion during the night, and a general decrease in water quality.

evaporation The process by which water becomes a vapor at a temperature below the boiling point. The rate of evaporation is generally expressed in inches or centimeters per day, month, or year.

evapotranspiration The evaporative loss of water from the earth's surface combined with losses from transpiration by plants and evaporation of water intercepted by plants and objects.

extension agents Government officials who serve as consultants to agriculture.

fecal coliform A bacterial group that is an indicator of human or animal pollution. See *indicator organisms.*

filterable residue test A test used to measure the total dissolved solids in water by first filtering out any undissolved solids and then evaporating the filtered water to dryness. The residue that remains is called *filterable residue* or *total dissolved solids.*

filtration The process of removing solids from water by passing the water through some material or device.

flotation A process for separating solids from water by using air to float the particles.

flow General term for movement of water, commonly used to mean (imprecisely) instantaneous flow rate, average flow rate, or volume.

flow rate The volume of water passing by a point per unit of time. Flow rates are either instantaneous or average.

frazil ice Small ice crystals that can block water intakes.

free residual chlorine Hypochlorous acid in water that responds to testing procedures and that serves as an active disinfectant.

geographic information system (GIS) A computer hardware and software system that captures, analyzes, and displays interrelated and geographically linked data.

Giardia lamblia A protozoan that can survive in water and that causes human disease.

granular activated carbon A chemical used to remove certain dissolved contaminants from water.

granular media A material used for filtering water, consisting of grains of sand or other material.

gravel pack Gravel surrounding the well intake screen, artificially placed ("packed") to aid the screen in filtering out the sand of an aquifer. Needed only in aquifers containing a large proportion of fine-grained material.

groundwater Subsurface water occupying the saturation zone, from which wells and springs are fed. In a strict sense, the term applies only to water below the water table. Contrast with *surface water.*

groundwater aquifer See *aquifer.*

groundwater classification A scheme to categorize groundwater according to its quality.

hardness A characteristic of water, caused primarily by the salts of calcium and magnesium. Causes deposition of scale in boilers, damage in some industrial processes, and sometimes objectionable taste. May also decrease the effectiveness of soap.

hazardous-waste facilities Depository areas where dangerous materials are stored.

hydraulic conductivity A measure of the ease with which water will flow through geologic formations.

hydraulic jetting The use of water forced through the well screen to suspend fine particles in well development.

hydraulics The branch of science that deals with fluids at rest and in motion.

hydric soils Soils that are either saturated or ponded long enough to be depleted of oxygen for extended periods. Hydric soils are found in wetlands and other water bodies.

hydrodynamics The study of water in motion.

hydrologic cycle The water cycle; the movement of water to and from the surface of the earth.

hydrology The science dealing with the properties, distribution, and circulation of water and its constituents in the atmosphere, on the earth's surface, and below the earth's surface.

hydrostatics The study of water at rest.

impermeable layer A layer not allowing, or allowing only with great difficulty, the movement of water.

impervious Resistant to the passage of water.

impoundment A pond, lake, tank, basin, or other space that is used for storage, regulation, and control of water.

indicator organisms A bacterium that does not cause disease but indicates that disease-causing bacteria may be present. See *fecal coliform*.

indirect potable reuse The use of water from streams that have upstream discharges of wastewater.

industrial water use Use of water in industrial activities such as power generation, steel manufacturing, pulp and paper processing, and food processing.

infiltration The flow or movement of water through soil.

infiltration gallery A subsurface structure to receive water filtered through a streambed.

inorganic material Chemical substances of mineral origin, or more correctly, not of basically carbon structure.

instantaneous flow rate The flow rate (volume per time) at any instant. Defined by the equation $Q = AV$, where Q = flow rate, A = cross-sectional area of the water, and V = velocity.

intake structure A structure or device placed in a surface water source to permit the withdrawal of water from that source.

interception The interruption of precipitation by an object on the earth's surface that limits it from reaching the ground surface; usually stems of vegetation and leaves in undeveloped watersheds, buildings and roads in developed watersheds.

ion exchange A process that removes undesirable chemicals from water and replaces them with harmless ones.

maximum contaminant level (MCL) The maximum permissible level of a contaminant in water as specified in the regulations of the Safe Drinking Water Act.

membrane processes Water purification processes using thin plastic films.

milligrams per liter (mg/L) A unit of the concentration of water or wastewater constituents. One mg/L is equivalent to 0.001 g of the constituent in 1,000 mL (1 L) of water. For reporting the results of water and wastewater analyses, this unit has replaced parts per million, to which it is approximately equivalent.

minimum day The day during which the minimum-day demand occurs.

minimum-day demand Least volume per day flowing through the plant for any day of the year.

minimum-hour demand Least volume per hour flowing through a plant for any hour in the year.

minimum-month demand Least volume of water passing through the plant during a calendar month.

molecule The smallest particle of a compound that has the compound's characteristics.

mrem Millirem. See *rem*.

National Pollution Discharge Elimination System (NPDES) A licensing system for waste discharges.

nephelometric turbidity unit (ntu) The unit of measure used to express the turbidity (cloudiness) of a water sample.

nitrogen fertilizer A plant nutrient that can cause algae blooms in sensitive water bodies.

nonpoint sources Material entering a water body that comes from overland flow or groundwater discharge rather than out of a pipe.

numerical modeling A computerized technique for determining an aquifer recharge area.

observation well Well placed near a production well to monitor changes in the aquifer.

organic adsorption The removal of volatile organic chemicals and synthetic organic chemicals by use of activated carbon.

organic material Chemicals that contain carbon and are associated with living matter.

overlay maps Transparent sheets containing data that are placed over the base map and indicate special features.

overlying use The land use that occurs on top of an aquifer.

oxidant A chemical that oxidizes other materials.

oxidize To chemically combine with oxygen, or in the case of bacteria or other organics, the combination of oxygen and organic material to form more stable organics or minerals.

oxygen depletion A state in which oxygen has been used up so that little or none is left.

ozone A form of oxygen that has very strong oxidizing powers.

palatable Pleasing to the taste.

parts per million (ppm) Unit of measure indicating the number of weight or volume units of a constituent present for each 1 million units of the solution or mixture. Formerly used to express the results of most water and wastewater analyses, but recently replaced by the use of milligrams per liter.

pathogens Disease-causing organisms.

peak day The day during which the peak-day demand occurs.

peak-day demand The greatest volume per day flowing through a plant for any day of the year.

peak-hour demand The greatest volume per hour flowing through a plant for any hour in the year.

peak-month demand The greatest volume of water passing through the plant during a calendar month.

percolation The downward movement or flow of water through the pores of soil once filtration has occurred.

perennial stream A stream that flows continuously at all seasons of a year and during dry as well as wet years. Such a stream is usually fed by groundwater, and its water surface generally stands at a lower level than that of the water table in the locality.

pH A measure of the acidic or alkaline nature of water. Values of pH range from 1 (most acidic) to 14 (most alkaline); pure water, which is neutral (neither acidic nor alkaline), has a pH of 7, the center of the range.

phosphorus fertilizer A plant nutrient that can cause algae blooms in sensitive water bodies.

photochemical oxidation Changes in chemicals caused by sunlight.

photosynthesis The process by which plants, using the chemical chlorophyll, convert the energy of the sun into food energy. Through photosynthesis, all plants and ultimately all animals (which feed on plants or other, plant-eating animals) obtain the energy of life from sunlight.

piezometric surface The surface that coincides with the static water level in an artesian aquifer.

point source Water potentially containing contaminants coming from a discharge pipe.

police power The ability to enforce laws.

pollution A condition created by the presence of harmful or objectionable material in water.

poorly buffered Not resisting changes in pH.

porosity An indication of the volume of space within a given amount of a material.

powdered activated carbon Carbon, added to water in powdered form, used to remove chemicals from water.

precipitation The process by which atmospheric moisture is discharged onto land or water surface. Precipitation, in the form of rain, snow, hail, or sleet, is usually expressed as depth for a day, month, or year, and designated as daily, monthly, or annual precipitation.

primacy The acceptance by states of the task of enforcing the SDWA.

priority use The assigning of water rights based on who has been using the water the longest.

pristine Pure and free from contamination.

protozoa Small single-celled animals including amoebae, ciliates, and flagellates.

public water use Use of water to support public facilities such as public buildings and offices, parks, golf courses, picnic grounds, campgrounds, and ornamental fountains.

quagga mussel A bivalve that multiplies rapidly in fresh water and can clog intake pipes.

qualified right A water rights term indicating that the right to water use is not absolute but must be shared with others.

radial well A very wide, relatively shallow caisson that has horizontally drilled wells with screen points at the bottom. Radial wells are large producers.

radioactive Referring to a material that has an unstable atomic nucleus, which spontaneously decays or disintegrates, producing radiation.

rainfall intensity The amount of rainfall occurring during a specified time. Rainfall intensity is usually expressed as inches or centimeters per hour.

reasonable use A water rights term indicating that the water use is acceptable in general terms.

recharge Addition of water to the groundwater supply from precipitation and by infiltration from surface streams, lakes, reservoirs, and snowmelt.

recovery method A procedure for aquifer evaluation that measures how quickly water levels return to normal after pumping.

rem A unit of measure of radioactive effects on humans.

reuse water Wastewater treated to make it useful.

riparian doctrine A water right that allows the owners of land abutting a stream or other natural body of water to use that water.

riparians Those whose property holdings are along the shores of a water body.

riparian zone The area contiguous to the streambank of a flowing water body or the shoreline of an open water body, like a lake or pond. This area may be an upland or wetland, but is important to maintaining high water quality in the water body.

rule of correlative rights A water rights rule specifying that rights are not absolute but depend also on the rights of others.

safe yield The maximum dependable water supply that can be withdrawn continuously from a surface water or groundwater supply during a period of years in which the driest period, or period of greatest deficiency in water supply, is likely to occur.

saltwater intrusion The invasion of an aquifer by saltwater because of overpumping of a well.

sand barrier A layer of gravel around the curb of a dug well.

sanitary survey A detailed inspection of a facility with a view to public health concerns.

saturated The condition of a material when it can absorb no more of a second material. Saturated soil has its void spaces completely filled with water, so any water added will run off and not soak in.

screens A device for excluding debris on a surface water intake and for excluding sand in a groundwater supply.

SDWA The Safe Drinking Water Act.

secondary maximum contaminant level (SMCL) A nonenforceable regulation on a non-health-related contaminant.

sedimentation A treatment process using gravity to remove coagulated particles.

seepage The slow movement of water through small cracks or pores of a material.

service outlet A device used for releasing water at a dam for downstream uses.

siltation The accumulation of silt (small soil particles between 0.00016 and 0.0024 in. [0.004 and 0.061 mm] in diameter) in an impoundment.

slide gate A device on a dam to control the release of water.

sole-source aquifer The single water supply available in an area; also a USEPA designation for water protection.

solubilization The process of making a material easier to dissolve.

specific-capacity method A testing method for determining the adequacy of an aquifer or well.

specific yield A measure of well yield per unit of drawdown.

spillway A device used to release water from a dam.

spring A location where groundwater emerges on the surface of the ground.

sterilization The removal of all life from water, as contrasted with *disinfection*.

streambed permeability Measure of the ability of the bottom of a watercourse to transmit water to and from the underground.

stream channel The course through which a stream flows.

supply management The implementation of techniques by a utility to save water.

surface runoff (1) That portion of the runoff of a drainage basin that has not percolated beneath the surface after precipitation. (2) The water that reaches a stream by traveling over the soil surface or by falling directly into the stream channels, including not only the large permanent streams but also the tiny rills and rivulets.

surface water All water on the surface, as distinguished from *groundwater*.

surger A device used to develop a well.

surging and bailing A method used to develop a well by alternately increasing and decreasing water pressure against the walls of the well to dislodge small particles.

synthetic organic chemicals (SOCs) Chemicals produced by humans that can be water contaminants.

temperature A physical characteristic of water. The temperature of water is normally measured on one of two scales: Fahrenheit or Celsius.

thermal stratification The layering of water as a result of temperature differences.

thermoelectric power Electricity produced by heat in which water is used for steam production.

threshold odor The minimum odor of a water sample that can just be detected after successive dilutions of odorless water.

threshold odor number (TON) A numerical designation of the intensity of odor in water.

time of travel The time required for groundwater recharge to percolate through the soil and get to the wellhead.

topographic map A map that shows the elevation of the land in a specified area.

transpiration The process by which water vapor is lost to the atmosphere from living plants.

trihalomethanes (THMs) Compounds of chlorine and organic matter that are generated during disinfection.

turbidity A physical characteristic of water making the water appear cloudy. The condition is caused by the presence of suspended matter.

turnover The vertical circulation of water in large water bodies caused by the mixing effects of temperature changes and wind.

ultraviolet irradiation Use of a portion of the light spectrum to disinfect drinking water.

unaccounted-for water Water use that does not go through meters and thus is not paid for (including water that is wasted).

viruses The smallest and simplest form of life. The many types of viruses reproduce themselves in a manner that causes infectious disease in some larger life forms, such as humans.

void spaces A pore or open space in rock or granular material that is not occupied by solid matter. It may be occupied by air, water, or other gaseous or liquid material.

volatile organic chemicals (VOCs) Organics that are produced by humans, are primarily solvents, and are found in water as contaminants.

waiver for filtration An exception granted to a state so that water from a source can be served without prior filtration.

water body A natural or artificial basin in which water is impounded.

waterborne disease A disease caused by a waterborne organism or toxic substance.

water conservation The reduction of water usage, accomplished primarily by preventing waste.

watercourse (1) A running stream of water. (2) A natural or artificial channel for the passage of water.

water rights A body of law that determines water ownership.

water-saving fixtures Devices and appliances specifically designed to have low water use.

water-saving kits Devices sometimes distributed by utilities to reduce customer water usage.

watershed See *drainage basin.*

watershed boundary The watershed divide (the high point on a watershed), which directs the flow of water in a certain direction.

watershed divide See *watershed boundary.*

watershed drainage area See *drainage basin.*

water table The upper surface of the zone of saturation closest to the ground surface.

water-table aquifer An aquifer confined only by a lower impermeable layer.

water-table well A well constructed in a water-table aquifer.

well casing See *casing.*

well development The process of removing particles from an aquifer to produce potable water.

well field placement The arranging of wells to achieve a desired output from an aquifer.

wellhead protection A process of controlling potential groundwater contamination for a specific groundwater source.

wellhead protection program (WHPP) A program required by the Safe Drinking Water Act to achieve wellhead protection.

wetland Areas that are inundated or saturated by surface or groundwater at a frequency and duration sufficient to support, and that under normal circumstances do support, a prevalence of vegetation typically adapted for life in saturated soil conditions, including swamps, marshes, bogs, and similar areas.

Xeriscaping The selection and use of trees, shrubs, and plants that require small amounts of water for landscaping.

zebra mussel A bivalve that is a serious pest in surface water because it blocks intake pipes.

zone of saturation The part of an aquifer that has all the water it can contain.

zoning A civil process that controls land use.

Index

NOTE: *f* indicates a figure; *t* indicates a table

D

E

F

G

H

R

S

T

U

V

W

X

Z